QUESTIONS OF
SCIENCE

qs

QUESTIONS OF SCIENCE is an international series of books published by Harvard University Press, Penguin Books, and Editions Odile Jacob. Written by scientists of international reputation, these books provide the nonspecialist reader with an understanding of scientific thought at the frontiers of our knowledge. The series will treat the enduring and exciting questions in fields that particularly inform our sense of place in the universe.

*Down House, Charles and Emma Darwin's home in Kent, where
the geologist Charles Lyell met with Darwin in April 1856 and
persuaded him that he should put an abstract of his views on
evolution into print.*

ONE LONG ARGUMENT

*Charles Darwin and the Genesis of
Modern Evolutionary Thought*

ERNST MAYR

HARVARD UNIVERSITY PRESS

Cambridge, Massachusetts · 1991

Design by Marianne Perlak

Library of Congress cataloging information is on last page of book.

*For Michael T. Ghiselin
and Frank J. Sulloway,
who have contributed so much
to our understanding of Darwin*

Preface

A MODERN EVOLUTIONIST turns to Darwin's work again and again. This is not surprising, since the roots of all our evolutionary thinking go back to Darwin. Our current controversies very often have as their starting point some vagueness in Darwin's writings or a question Darwin was unable to answer owing to the insufficient biological knowledge available in his time. But one returns to Darwin's original writings for more than historical reasons. Darwin frequently understood things far more clearly than both his supporters and his opponents, including those of the present day.

An analysis of almost any scientific problem leads automatically to a study of its history. The many unresolved issues in evolutionary biology are no exception to this rule. To understand the history of a scientific problem, however, one must appreciate not only the state of factual knowledge but also the *Zeitgeist* of the time. Any investigator's interpretation of his observations or experiments depends mainly on this conceptual framework. For many years a major objective of my historical studies has been to discover the concepts—or sometimes, even more broadly, the ideologies—on which the theorizing of certain historical figures was based.

My interest in Darwin's thought arose in my university years, but my more active preoccupation began in 1959 with the centenary of the publication of *On the Origin of Species*. I studied Darwin's writings even more intensely when I prepared an introduction to a facsimile of the first edition of the *Origin*, published in 1964. Curiously, most editions of this work available before that time were of the much-revised sixth edition. The inexpensive facsimile made wide distribution of the first edition of the *Origin* possible for the first time since its original publication.

In the ensuing years I devoted myself to the study of Darwin's work, and this effort culminated in *The Growth of Biological Thought: Diversity, Evolution, and Inheritance,* published in 1982. However, in this overview of the history of systematics, evolutionary biology, and genetics it was impossible to present detailed analyses of certain aspects of Darwin's work. These I treated in separate lectures and papers, mostly presented at commemorative celebrations. Essays growing out of these lectures, along with many other articles I had written on the history and philosophy of biology, were collected in a volume entitled *Toward a New Philosophy of Biology: Observations of an Evolutionist* (1988).

While I was reviewing this rather technical and specialized collection for a German translation, it occurred to me that a separate volume devoted exclusively to Darwin and Darwinism might be useful to students and lay people broadly interested in the role of Darwin's thought in the history of ideas. As my starting point for this new volume, I pulled from my collected essays the chapters that were devoted principally to Darwin and Darwinism and began to edit them. Yet after judiciously pruning and shaping the original eight essays, I realized that considerable gaps remained in my presentation. I set about writing several new chapters, while thoroughly reworking and rearranging material from the original essays. The book that has resulted represents, I hope, a mature reflection on Darwin's thought—one that emphasizes previously neglected aspects of his work and clarifies controversial or confused issues.

The last thirty years have been a period of unparalleled activity in Darwin research, primarily owing to the discovery of the "Darwin papers"—notebooks, letters, unpublished manuscripts, and so on. The first six volumes of Darwin's letters have now been published, as has a single-volume edition of Darwin's notebooks. In addition to these primary documents, books about Darwin and his life appear every few years, and several of them are very good. Yet even today much that is written about Darwin is simply wrong or, worse, malicious—in large part because the author has failed to understand the concepts that underlie Darwin's thought and its development, and the entrenched ideologies that his "one long argument" (a phrase he used in Chapter 14 to describe the *Origin*) was designed to oppose. This volume is an attempt to correct some of those conceptual misunderstandings, as well as to incorporate many of the latest findings of the still exceedingly active Darwin research program.

The facts of evolution as well as particular problems of phylogeny are hardly mentioned in this book. It is of little relevance for evolutionary theory whether the ancestors of the molluscs were metameric (segmented) or not (almost surely they were), whether the coelenterates have the same ancestry as the flatworms, whether the tetrapods arose from lungfishes (as now seems more probable) or coelacanths. An enormous literature already exists on these concrete problems of phylogeny. Instead, I have concentrated on the mechanisms of evolution and on the historical development, from Darwin on, of the major concepts and theories of evolutionary biology.

Finally, my attention to underlying concepts in this book represents an effort to combat a disturbing trend in our modern view of basic scientific research. Science, in the minds of too many scientists, is considered to be merely a sequence of discoveries or, worse, a steppingstone to technological innovation. In my writings I hope to have established a better balance in the evaluation of science by showing the close connection between the questions raised within certain fields of science and other more general aspects of modern thought and inquiry. This endeavor automatically leads to a consideration of Darwin, for no one has influenced our modern worldview—both within and beyond science—to a greater extent than has this extraordinary Victorian. We turn to his work again and again, because as a bold and intelligent thinker he raised some of the most profound questions about our origins that have ever been asked, and as a devoted and innovative scientist he provided brilliant, often world-shattering answers.

Contents

1 · Who Is Darwin? *1*

2 · Confronting the Creationists: The First
Darwinian Revolution *12*

3 · How Species Originate *26*

4 · Ideological Opposition to Darwin's Five Theories *35*

5 · The Struggle against Physicists and Philosophers *48*

6 · Darwin's Path to the Theory of Natural Selection *68*

7 · What Is Darwinism? *90*

8 · A Hard Look at Soft Inheritance: Neo-Darwinism *108*

9 · Geneticists and Naturalists Reach a Consensus:
The Second Darwinian Revolution *132*

10 · New Frontiers in Evolutionary Biology *141*

References *167*

Glossary *177*

Acknowledgments *189*

Index *191*

Illustrations

Frontispiece: Down House, Darwin's home in Kent. (Reproduced by permission of Mary Evans Picture Library, London.)

Following page 80:

J. S. Henslow (1796–1861), Darwin's botany professor at Cambridge University. (Reproduced by permission of the Syndics of the Fitzwilliam Museum, Cambridge.)

The *H.M.S. Beagle,* in the Straits of Magellan, 1833. (Reproduced by permission of the National Maritime Museum, Greenwich.)

The route of the voyage of the *Beagle,* 1831–1836. (By Laszlo Meszoly.)

The naturalist Jean Baptiste Lamarck (1744–1829). (Reproduced by permission of Bibliothèque centrale du Muséum National d'Histoire Naturelle, Paris.)

Charles Lyell (1797–1875), geologist

The ornithologist John Gould (1804–1881). (Reproduced by permission of the Ipswich Borough Museums and Galleries, Suffolk.)

Emma Wedgwood Darwin (1808–1896) in 1840, Charles Darwin's wife

Charles Darwin in 1840

The botanist Joseph Dalton Hooker (1817–1911), Darwin's friend and supporter. (Reproduced by permission of The Royal Society, London.)

The morphologist and paleontologist Richard Owen (1804–1892), who viciously attacked the *Origin*. (Reproduced by permission of the President and Council of the Royal College of Surgeons of England.)

Alfred Russel Wallace (1823–1913), co-discoverer with Darwin of the theory of evolution through natural selection

The morphologist, physiologist, and embryologist Thomas Henry Huxley (1825–1895), "Darwin's bulldog." (Reproduced by permission of Lady Huxley.)

The botanist Asa Gray (1810–1888), Darwin's most important supporter in America

The Swiss-born naturalist Louis Agassiz (1809–1873). (Reproduced by permission of the Museum of Comparative Zoology, Harvard University.)

The biologist Ernst Haeckel (1834–1919), an enthusiastic supporter and popularizer of Darwinism in Germany. (Courtesy of the National Library of Medicine, Bethesda, Maryland.)

August Weismann (1834–1914), the nineteenth century's greatest evolutionist after Darwin. (Reproduced by permission of the Museum of Comparative Zoology, Harvard University.)

Charles Darwin on the veranda at Down House. (Reproduced by permission of the New York Academy of Sciences. Courtesy of the Museum of Comparative Zoology, Harvard University.)

ONE LONG ARGUMENT

CHAPTER ONE

Who Is Darwin?

H ISTORICAL PERIODS are dominated by distinct sets of ideas which, taken together, form a well-defined *Zeitgeist*. Greek philosophy, Christianity, Renaissance thought, the Scientific Revolution, and the Enlightenment are examples of sets of ideas that dominated their historical period. The changes from one period to the next are usually rather gradual; other changes— more abrupt—are often referred to as revolutions. The most far-reaching of all these intellectual upheavals was the Darwinian revolution. The worldview formed by any thinking person in the Western world after 1859, when *On the Origin of Species* was published, was by necessity quite different from a worldview formed prior to 1859. It is almost impossible for a modern person to project back to the early half of the nineteenth century and reconstruct the thinking of this pre-Darwinian period, so great has been the impact of Darwinism on our views.

The intellectual revolution generated by Darwin went far beyond the confines of biology, causing the overthrow of some of the most basic beliefs of his age. For example, Darwin refuted the belief in the individual creation of each species, establishing in its place the concept that all of life descended from a common ancestor. By extension, he introduced the idea that humans were not the special products of creation but evolved according to principles that operate everywhere else in the living world. Darwin upset current notions of a perfectly designed, benign natural world and substituted in their place the concept of a struggle for survival. Victorian notions of progress and perfectibility were seriously undermined by Darwin's demonstration that evolution brings about change and adaptation, but it does not necessarily lead to progress, and it never leads to perfection.

Furthermore, Darwin established the basis for entirely new approaches in philosophy. At a time when the philosophy of science was dominated by a methodology based on mathematical principles, physical laws, and determinism, Darwin introduced the concepts of probability, chance, and uniqueness into scientific discourse. His work embodied the principle that observation and the making of hypotheses are as important to the advancement of knowledge as experimentation.

Darwin would be remembered as an outstanding scientist even if he had never written a word about evolution. Indeed, the evolutionist J. B. S. Haldane went so far as to say that Darwin's most original contribution to biology was not the theory of evolution but his series of books on experimental botany published nearer the end of his life (Haldane 1959:358). This achievement is little known among nonbiologists, and the same is true for his equally outstanding work on the adaptation of flowers and on animal psychology, as well as his competent work on the barnacles and his imaginative work on earthworms. In all these areas Darwin was a pioneer, and although in some areas more than a half a century passed before others built on the foundations he laid, it is now clear that Darwin attacked important problems with extraordinary originality, thereby becoming the founder of several now well-recognized separate disciplines (Ghiselin 1969). Darwin was the first person to work out a sound theory of classification, one which is still adopted by the majority of taxonomists. His approach to biogeography, in which so much emphasis was placed on the behavior and the ecology of organisms as factors of distribution, is much closer to modern biogeography than the purely descriptive-geographical approach that dominated biogeography for more than a half-century after Darwin's death.

Who was this extraordinary man, and how did he come to his ideas? Was it his training, his personality, his industry, or his genius that accounts for his success? Indeed, as we shall see, all were involved.

The Man and His Work

Charles Darwin was born on February 12, 1809, at Shrewsbury, England, the fifth of six children and the second son of Dr. Robert Darwin, an eminently successful physician. His grandfather was Erasmus Darwin, the author of *Zoonomia,* a work which anticipated his grandson's evolutionary interests by attempting to explain organic life according to evolutionary principles. His mother, the daughter of Josiah Wedgwood, the celebrated potter, died when Charles was only eight years old, and his elder sisters tried to fill her place.

From his earliest youth, Darwin was a passionate lover of the outdoors. As he himself said, "I was born a naturalist." Every aspect of nature intrigued him. He loved to collect, to fish and hunt, and to read nature books. Shrewsbury was a country town of about 20,000 inhabitants—a perfect place for the development of a naturalist, much better than either a big city or a strictly rural area would be.

School, consisting largely of the study of the classics, bored the young naturalist intolerably. Before he turned seventeen years old, Darwin's father sent him to the University of Edinburgh to study medicine like his older brother. But medicine appalled Charles, and he continued to devote much of his time to the study of nature. When it became clear that he did not want to become a physician, his father sent him early in 1828 to Cambridge to study theology. This seemed a reasonable choice, since virtually all the naturalists in England at that time were ordained ministers, as were the professors at Cambridge who taught botany (J. S. Henslow) and geology (Adam Sedgwick). Darwin's letters and biographical notes give the impression that in Cambridge he devoted more time to collecting beetles, discussing botany and geology with his professors, and hunting and riding with similarly inclined friends than to his prescribed studies. Yet he did well in his examinations, and when he took his B.A. in 1831 he stood tenth on the list of nonhonors students. More importantly, when Darwin had completed his Cambridge years he was an accomplished young naturalist.

Immediately upon finishing his studies Darwin received an invitation to join *H. M. S. Beagle* as naturalist and gentleman companion of Captain Robert FitzRoy. FitzRoy had been commissioned to survey the coasts of Patagonia, Tierra del Fuego, Chile, and Peru to provide information for making better charts. The voyage was to be completed within two or three years but actually lasted five. The *Beagle* left Plymouth on December 27, 1831, when Darwin was twenty-two years old, and returned to England on October 2, 1836. Darwin used these five years to their fullest extent. In an eminently readable travelogue (*Journal of Researches*) he tells about all the places he visited—volcanic and coral islands, tropical forests in Brazil, the vast pampas of Patagonia, a crossing of the Andes from Chile to Tucuman in Argentina, and much, much more. Every day brought unforgettable new experiences, an invaluable background for his life's work. He collected specimens from widely different groups of organisms, he dug out important fossils in Patagonia, he devoted much of his time to geology, but most of all he observed aspects of nature and asked himself innumerable questions as to the how and why of natural processes. He asked "why" questions not only about geological features and animal life, but also about political and social situations. And it was his ability to ask profound questions and his perseverance in trying to answer them that would eventually make Darwin a great scientist.

In spite of being desperately seasick every time the ship encountered rough weather, Darwin managed to read a great deal of important scientific literature that he brought along on the voyage. No scientific work was more critical to his further thinking than the first two volumes of Charles Lyell's *Principles of Geology* (1832), which not only gave Darwin an advanced course in uniformitarian geology—a theory that changes in the earth's surface have occurred gradually over long periods of time—but also introduced him to Jean Baptiste Lamarck's arguments for, and Lyell's arguments against, evolutionary thinking.

When Darwin boarded the *Beagle* he still believed in the fixity of species, as did Lyell and all of his teachers at Cambridge. Yet during the South American phase of the *Beagle* voyage Darwin made many observations that greatly puzzled him and that shook his belief in the fixity of species. But it was really his visit to the Galapagos in September and October 1835 that provided him with the crucial evidence, even though—being preoccupied during his stay with geological researches—he did not at first realize it. However, nine months later, in July 1836, he penned these words in his diary: "When I see these islands in sight of each other and possessed of but a scanty stock of animals, tenanted by these birds but slightly differing in structure and filling the same place in nature, I must suspect they are varieties . . . if there is the slightest foundation for these remarks, the zoology of the archipelagoes will be well worth examining: for such facts would undermine the stability of species" (Barlow 1963).

After his arrival in England in October 1836 Darwin sorted his collections and sent them to various specialists to be described in the official account of the *Beagle* expedition. In March 1837, when the celebrated ornithologist John Gould insisted that the mockingbirds (*Mimus*) collected by Darwin on three different islands in the Galapagos were three distinct species rather than varieties, as Darwin had thought, Darwin first understood the process of geographic speciation (Sulloway 1982; 1984): that a new species can develop when a population becomes geographically isolated from its parental species. Furthermore, if colonists derived from a single South American ancestor could become three species in the Galapagos, then all the species of mockingbirds on the mainland could have been derived from an ancestral species, and so could have, at an earlier time, the species of related genera, and so forth. Numerous statements in Darwin's writings confirm that from the spring of 1837 on he firmly believed in the gradual origin of new species through geographic speciation, and in the theory of evolution by common descent (see Chapter 2). But another year and a half would pass before Darwin figured out the mechanism of evolution, the principle of natural selection. This happened on September 28, 1838, as he was reading Malthus's *Essay on the Principle of Population* (see Chapter 6).

In January 1839 Darwin married his cousin Emma Wedgwood, and in September 1842 the young couple moved from London to a country house in the small village of Down (Kent), sixteen miles south of London, where Darwin lived until he died on April 19, 1882. Darwin's health required the move to a quiet place in the country. After he had passed his thirtieth year there were often long periods when he was unable to work more than two or three hours a day—indeed, when he was completely incapacitated for months on end. The exact nature of his illness is still controversial, but all the symptoms indicate a malfunctioning of the autonomous nervous system.

On the Origin of Species

Darwin did not publish his theories about evolution for another twenty years, even though he wrote some preliminary manuscript essays in 1842 and 1844. He devoted these years to his geological books and papers and to his monumental two-volume monograph on the barnacles (*Cirripedia*). Why did Darwin spend eight years on this piece of taxonomy instead of rushing into print with his important discovery of evolution by common descent through natural selection? Modern historical researches by Ghiselin (1969) and others have clearly shown that the barnacle studies were, for Darwin, an advanced graduate course in taxonomy, morphology, and ontogenetic research and not at all a waste of time. The experience he gained in these researches was an invaluable preparation for writing the *Origin*.

Finally, in April 1856, Darwin began to compose what he considered to be his "big species book." About two years later, after he had finished the first nine or ten chapters of this book, he received a letter from the naturalist Alfred Russel Wallace, who at that time was collecting specimens in the Moluccas. This letter, which Darwin received in June 1858, was accompanied by a manuscript that Wallace asked Darwin to read and submit to some journal if he found it acceptable. When Darwin read the manuscript, he was thunderstruck. Wallace had arrived at essentially the same theory of evolution by common descent through natural selection as he. On July 1, 1858, Darwin's friends Charles Lyell and the botanist Joseph Hooker presented Wallace's manuscript, together with excerpts from Darwin's manuscripts and letters, at a meeting of the Linnean Society of London. This presentation amounted to a simultaneous publication of the findings of Darwin and Wallace. Darwin quickly abandoned his idea of finishing his monumental work on the species and wrote instead what he called an "abstract," which became his famous *On the Origin of the Species,* published November 24, 1859.

The impact of the *Origin* was enormous. Quite rightly it has been referred to as "the book that shook the world." In its first year the work sold 3,800 copies, and in Darwin's lifetime the British printings alone sold more than 27,000 copies. Several American printings, as well as innumerable translations, also appeared. Nevertheless, only in our lifetime have historians understood how fundamental the influence of this work has been. Every modern discussion of man's future, the population explosion, the struggle for existence, the purpose of man and the universe, and man's place in nature rests on Darwin.

In the ensuing twenty-three years of his life Darwin worked steadily on certain aspects of evolution he had not been able to cover adequately in the *Origin.* In a two-volume work, *The Variation of Animals and Plants under Domestication* (1868), he struggled valiantly with the problem of how genetic variation originates. In *The Descent of Man and Selection in Relation to Sex* (1871) he dealt with the evolution of the human species and expanded on his theory of sexual selection. *The Expression of the Emotions in Man and Animals* (1872) laid a foundation for the study of animal behavior. *Insectivorous Plants* (1875) described the remarkable adaptation of the sun-dew and other plants for catching and digesting insects. In *The Effects of Cross- and Self-Fertilization in the Vegetable Kingdom* (1876), in *The Different Forms of Flowers on Plants of the Same Species* (1877), and in *The Power of Movement in Plants* (1880), Darwin discussed aspects of plant growth and physiology, as indicated in the titles. And finally in *The Formation of Vegetable Mold, through the Action of Worms, with Observations on Their Habits* (1881), he described the important role played by earthworms in the formation of the topsoil.

How could one man achieve so much in a lifetime, particularly considering the constraints imposed by his illness ? Only by retreating into the quiet of the countryside, refusing to accept most offered offices or memberships on committees, and, through the generosity of his father, living on his inherited income was Darwin able to complete his task. Yet Darwin was not a recluse. He kept in touch with the scientific world through an extensive correspondence and occasional visits to London, and he was a devoted husband and dedicated father to his ten children.

Darwin was described by his contemporaries as an extraordinarily modest, gentle person who went out of his way to avoid hurting anyone's feelings. He worked so hard because he had an unquenchable thirst for learning, not in order to get advancement or honors. In his publications he was a scientist's scientist. He did not write for the general public; when some of his works had great popular success, he was always astonished.

Nevertheless, Darwin fought for the recognition of his findings among scientists, and he was supported by a small band of loyal friends—among them Lyell, Hooker, and the morphologist T. H. Huxley, often referred to as Darwin's bulldog because it was he who in public debates most often defended Darwin's theories. The most fervent of Darwin's admirers were the naturalists. These included the codiscoverer of evolution by natural selection, A. R. Wallace, the entomologist Henry Walter Bates, and the naturalist Fritz Müller.

To have a loyal group of defenders was very important because Darwin was attacked with unusual ferocity. In 1860 the Harvard University zoologist Louis Agassiz wrote that Darwin's theory was a "scientific mistake, untrue in its facts, unscientific in its methods, and mischievous in its tendency." The *Origin* was extensively reviewed in journals by the leading philosophers, theologians, literary men, and scientists of the day. By far the majority of the reviews were negative, if not extremely hostile (Hull 1973). Curiously, this negative reception continued after Darwin's death in 1882 and has lasted in certain circles to the present day.

Darwin's Scientific Method

The years during which Darwin worked on the manuscript for his big species book happened to be the same years in which the field of study known as the philosophy of science originated in England. While still a student, Darwin had read with enthusiasm John Herschel's *Preliminary Discourse on the Study of Natural Philosophy* (1830), and this work continued to be one of his favorite readings. He also read William Whewell's and John Stuart Mills' books and tried conscientiously to follow their prescriptions for the study of natural history (Ruse 1970; Hodge 1982). This was rather difficult, since the recommendations of the various authors were often contradictory; as a result, so were Darwin's own statements on the subject. To satisfy some of his readers Darwin asserted that he followed "the true Baconian method" (Darwin 1958:119), that is, straight induction. In reality, he "speculated" on any subject he encountered. He realized that one cannot make observations unless one has some hypothesis on the basis of which to make the appropriate observations. Therefore, "I can have no doubt that speculative men, with a curb on, make far the best observers" (Darwin 1988:317). He stated his views most clearly in a letter to Henry Fawcett. "About 30 years ago there was much talk that geologists ought to observe and not to theorize; and I well remember someone saying that at this rate a man might as well go into a gravel pit and count the pebbles and describe the colors. How odd it is that anyone should not see that all observation must be for or against some view if it is to be of any service!" (Darwin and Seward 1903).

Darwin's method was actually the time-honored method of the

best naturalists. They observe numerous phenomena and always try to understand the how and why of their observations. When something does not at once fall into place, they make a conjecture and test it by additional observations, leading either to a refutation or strengthening of the original assumption. This procedure does not fit well into the classical prescriptions of the philosophy of science, because it consists of continually going back and forth between making observations, posing questions, establishing hypotheses or models, testing them by making further observations, and so forth. Darwin's speculation was a well-disciplined process, used by him, as by every modern scientist, to give direction to the planning of experiments and to the collecting of further observations. I know of no forerunner of Darwin who used this method as consistently and with as much success.

That Darwin was a genius is hardly any longer questioned, some of his earlier detractors notwithstanding. But there must have been a score of other biologists of equal intelligence who failed to match Darwin's achievement. What is it that distinguishes Darwin from all the others? Perhaps we can answer this question by investigating what kind of scientist Darwin was. As he has said, he was first and foremost a naturalist. He was a splendid observer, and like all other naturalists he was interested in organic diversity and in adaptation. Naturalists are, on the whole, describers and particularizers, but Darwin was also a great theoretician, something only very few naturalists have ever been. In that respect Darwin resembles much more some of the leading physical scientists of his day. But Darwin differed from the run-of-the-mill naturalists also in another way. He was not only an observer but also a gifted and indefatigable experimenter whenever he dealt with a problem whose solution could be advanced by an experiment.

I think this suggests some of the sources of Darwin's greatness. The universality of his talents and interests had preadapted him to become a bridge-builder between fields. It enabled him to use his background as a naturalist to theorize about some of the most challenging problems that pique our curiosity. And, in the face of widespread beliefs to the contrary, Darwin was utterly bold in his theorizing. A brilliant mind, great intellectual boldness, and an ability to combine the best qualities of a naturalist-observer, philosophical theoretician, and experimentalist—the world has so far seen such a combination only once, and it was in the man Charles Darwin.

CHAPTER TWO

Confronting the Creationists:
The First Darwinian Revolution

A NEW DISCOVERY IN SCIENCE, such as the double-helix structure of DNA, is usually accepted almost immediately. If the assumed discovery turns out to be based on an error or a misinterpretation, it quickly disappears from the literature. By contrast, resistance to the introduction of new theories, particularly those that are based on new concepts, is much stronger and broader-based. Isaac Newton's theory that gravitation is responsible for the motion of the planets required some eighty years before it was universally accepted. Alfred Wegener's theory of continental drift was published in 1912 but was generally adopted only fifty years later, after the acceptance of the theory of plate tectonics.

Theories that either implied or overtly assumed organic evolution had been proposed from Buffon (1749) on, most explicitly by Lamarck (1809), but were largely ignored or actively resisted. When Charles Darwin first began to think about such problems, in his Cambridge days and on the *Beagle,* all of his teachers and friends firmly believed that species do not change. They held this belief in large part because of their religious views. The two teachers in Cambridge to whom Darwin was closest, Henslow and Sedgwick, were both orthodox Christians and accepted the dogma of the Bible literally, including the story of creation. Even the geologist Charles Lyell, whose work profoundly influenced Darwin's thinking—although Darwin did not meet him personally until after he returned from the *Beagle* voyage—was a theist who believed that species were created by God's hand. In all the writings of the naturalists, geologists, and philosophers of the period, God played a dominant role. They saw nothing peculiar in explaining otherwise puzzling phenomena as being caused by God, and that included the question of how species originate.

The Argument against Creationism

When Darwin decided in 1827 to study for the ministry, he too was an orthodox Christian. As he said in his autobiography (1958:57), "As I did not then in the least doubt the strict and literal truth of every word in the Bible," he neither questioned the occurrence of miracles nor any other supernatural phenomena. When he sailed on the *Beagle,* he reports, "I was quite orthodox, and I remember being heartily laughed at by several of the officers for quoting the Bible as an unanswerable authority on some point of morality" (1958:85). And yet it is evident that many of the experiences he had during the five years of his voyage raised the first doubts in his mind about his religious beliefs. How could a wise and good Creator permit the unspeakable cruelty and sufferings of slavery? How could he instigate earthquakes and volcanic eruptions that killed thousands or tens of thousands of innocent people? Yet Darwin was far too busy with his work to become obsessed by such disturbing thoughts. After his return from the *Beagle* Darwin was more strongly influenced by his family's beliefs than by his Cambridge friends. As Ruse (1979:181) has observed: "His grandfather Erasmus was at best a weak deist, quite able to believe in evolution . . . his father Robert, who had an overwhelming influence on Darwin, was an unbeliever; his uncle Josiah Wedgwood was a unitarian; and most important of all, Charles's older brother Erasmus had become an unbeliever by the time Charles returned from the Beagle voyage."

But a more important influence on his changing beliefs than his intellectual environment was Darwin's own scientific findings. Almost everything he learned in his natural-history studies was more or less in conflict with Christian dogma. Every species had numerous adaptations, from species-specific songs or courtships to specialized food and specific enemies. According to the philosophy of natural theology, which was widely accepted in England at that time, God had designed and looked after all of these numberless details. They could not possibly be controlled by physical laws, for these details were far too specific. Laws can control the physical world, where adaptation was absent, but the specificities and adaptations of the organic world required that God personally look after every detail relating to thousands or, as we now know, millions of species. Darwin could not accept this explanation of the enormous diversity and adaptation he observed, and he found himself more and more inclined toward natural mechanisms.

Darwin's observations were also in conflict with the natural theologians' belief in a perfect world. What the naturalist finds instead are numerous imperfections. How could all the species of former periods have become extinct if they had been perfect? As Hull (1973:126) has said rightly, "The god implied by . . . a realistic appraisal of the organic world was capricious, cruel, arbitrary, wasteful, careless, and totally unconcerned with the welfare of his creations." Such considerations as these gradually drove Darwin to the decision to try to explain the world without invoking God or any kind of supernatural forces.

As we will see in more detail below, Darwin adopted natural selection in the fall of 1838, which would suggest that he had decided to reject supernatural explanations prior to that date. But this conclusion has been vigorously denied by several authors, whose views are seemingly well supported by Darwin's own statements in the *Origin* and elsewhere. There is now in the Darwin literature an extreme spread of opinions, ranging from the conclusion that Darwin was already an agnostic in 1837 when he began to write his *Notebooks*, to the notion that he was still a theist in 1859 and became an agnostic only late in life. How could such a wide divergence of interpretation develop in view of the massive material relating to the issue in Darwin's own notes, letters, and publications? The answer is Darwin's own ambiguity. This has been splendidly analyzed by Kohn (1989). Additional light on changes in Darwin's thinking has been provided by Moore (1989). Reading these accounts (as well as earlier ones by Ospovat, Greene, Gillespie, Moore, Manier, and Richards) leads me to the conclusions that follow. However, Kohn has rightly said, every interpreter of Darwin's religiosity has tended to read into Darwin what he wanted to find, and presumably I am not escaping this weakness either.

It is quite evident that prior to the end of July 1838 Darwin had made quite a few notebook entries that were thoroughly "materialistic" (= agnostic). But July 29–31, 1838, Darwin visited the Wedgwoods at Maer in Staffordshire, and the courtship of Charles and his cousin Emma began with that visit. Emma was an orthodox Christian and, as has often been pointed out, it became clear to Darwin that it would destroy their marriage if he was not cautious in the expression of his religious views. More than that, as Kohn has said quite rightly, "Emma became Darwin's model of the conventional Victorian reader" (1989:226). She clearly had an effect on "the contruction of Darwin's texts. To me this means one crucial thing, not a word of the ambiguous God-talk of the *Origin* can be taken at face value" (Kohn 1989:226). To be sure, it seems that Darwin himself was still wavering. "Atheism both attracted and frightened him" (p. 227). He was aware of the great unknown, and it would have been a comfort to him to believe in a supreme being. But all the phenomena of nature he encountered were consistent with a straightforward scientific explanation that did not invoke any supernatural agencies.

Thirteen years later, in 1851, an event occurred in Darwin's life that thoroughly affected him. He lost his beloved ten-year-old daughter Annie, a child seemingly perfect in her goodness. As Moore (1989) describes, this "cruel" event seems to have extinguished the last traces of theism in Darwin.

Whether one wants to call him a deist, an agnostic, or an atheist, this much is clear, that in the *Origin* Darwin no longer required God as an explanatory factor. Creation as described in the Bible was contradicted for Darwin by almost every aspect of the natural world. Furthermore, creation simply could not explain the fossil record, nor the hierarchy in types of organisms that had been proposed by the taxonomist Carl Linnaeus, nor many of the other findings of science. Yet almost all of Darwin's peers still believed in some form of creation, and many of Darwin's contemporaries accepted Bishop Ussher's calculation that creation had occurred as recently as 4004 years B.C.

By contrast, the geologists had long been aware of the immense age of the earth, which would have allowed plenty of time for abundant organic evolution. Another discovery of geology that was most important and most disturbing for the creationist was the discovery of abundant extinction. Already in the eighteenth century the German naturalist Johann Friedrich Blumenbach and others had accepted extinction of formerly existing types like ammonites, belemnites, and trilobites, and of entire faunas, but it was not until Cuvier worked out the extinction of a whole sequence of mammalian faunas in the Tertiary of the Paris Basin that the acceptance of extinction became inevitable. The ultimate proof for it was the discovery of fossil mastodons and mammoths, animals so huge that any living survivors could not possibly have remained undiscovered in some remote part of the globe.

Three explanations for extinction were offered. According to Lamarck, no organism ever became extinct; there simply was such drastic transformation that formerly existing types had changed beyond recognition. According to another school, one to which Louis Agassiz belonged, each former fauna had become extinct as a whole through some catastrophe and was replaced by a newly created, more progressive fauna. This had happened, according to Agassiz, fifty times since the earth was formed. Such catastrophism was unpalatable to Lyell, who produced a third theory consistent with his uniformitarianism. He believed that individual species became extinct one by one as conditions changed and that the gaps thus created in nature were filled by the introduction of new species through some presumably supernatural means. Lyell's theory was an attempt at a reconciliation between those who recognized a changing world of long duration and those who supported the tenets of creationism.

The question of precisely how these new species were introduced was left unanswered by Lyell. He bequeathed this problem to Darwin, who in due time made it his most important research program. Darwin thus approached the problem of evolution in an entirely different manner from Lamarck. For Lamarck, evolution was a strictly vertical phenomenon, proceeding in a single dimension, that of time. Evolution for him was a movement from less perfect to more perfect, from the most primitive infusorians up to the mammals and man. Lamarck's *Philosophie Zoologique* was the paradigm of vertical evolutionism. Species played no role in Lamarck's thinking. New species originated all the time by spontaneous generation from inanimate matter, but this produced only the simplest infusorians. Each newly established evolutionary line gradually moved up to ever greater perfection, as organisms adapted to their environment and passed along to their offspring these newly acquired traits.

Darwin was unable to build on this foundation but rather started from the fundamental question that Lyell had bequeathed to him, namely, how do new species originate? Although Lyell had appealed to "intermediate causes" as the source of the new species, the process was nevertheless a form of special creation. "Species may have been created in succession at such times and at such places as to enable them to multiply and endure for an appointed period and occupy an appointed space on the globe" (Lyell 1835, 3:99–100). For Lyell, each creation was a carefully planned event. The reason why Lyell, like Henslow, Sedgwick, and all the others of Darwin's scientific friends and correspondents in the middle of the 1830s, accepted the unalterable constancy of species was ultimately a philosophical one. The constancy of species—that is, the inability of a species, once created, to change—was the one piece of the old dogma of a created world that remained inviolate after the concepts of the recency and constancy of the physical world had been abandoned.

Evidence for the Gradual Evolution and Multiplication of Species

No genuine and testable theory of evolution could develop until the possibility was recognized that species have the capacity to change, to become transformed into new species, and to multiply into several species. For Darwin to accept this possibility required a fundamental break with Lyell's thinking. The question which we must ask ourselves is how Darwin was able to emancipate himself from Lyell's thinking, and what observations or conceptual changes permitted Darwin to adopt the theory of a transforming capacity of species.

As Darwin tells us in his autobiography, he encountered many phenomena during his visit to South America on the *Beagle* that any modern biologist would unhesitatingly explain as clear evidence for evolution. Yet even when sorting his collections on the homeward voyage, and realizing that the "varieties" he observed "would undermine the stability of species," Darwin at that date (approximately July 1836) had not yet consciously abandoned the concept of constant species (Barlow 1963). This Darwin apparently did in two stages. The discovery of a second, smaller species of rhea (South American ostrich) led him to the theory that an existing species could give rise to a new species, by a sudden leap or saltation. Such an origin of new species had been postulated scores of times before, from the Greeks to Robinet and Maupertuis (Osborn 1894). Sudden new origins, however, are not evolution. The diagnostic criterion of evolutionary transformation is gradualness.

The concept of gradualism, the second step in Darwin's conversion, was apparently first adopted by Darwin when the ornithologist John Gould, who prepared the scientific report on Darwin's bird collections, pointed out to him that there were three different endemic species of mockingbirds on three different islands in the Galapagos. Darwin had thought they were only varieties (Sulloway 1982b). The mockingbird episode was of particular importance to Darwin for two reasons. The Galapagos endemics were quite similar to a species of mockingbirds on the South American mainland and clearly derived from it. Thus, the Galapagos birds were not the result of a single saltation, as Darwin had postulated for the new species of rhea in Patagonia, but had gradually evolved into three separate but similar species on three different islands. This fact helped to convert Darwin to the concept of gradual evolution (see Chapter 3). Even more important was the fact that these three different species had branched off from a single parental species, the mainland mockingbird—an observation that gave Darwin a solution to the problem of speciation—that is, how and why species multiply.

For a believer in saltational evolution, speciation meant the sudden change of a species into a different one. For a believer in transformational evolutional, speciation meant the gradual change of a species into a different one in a phyletic lineage. But neither of these two theories explained the origin of the enormous organic diversity Darwin saw around him. There are at least ten million species of animals and almost two million species of plants on the Earth today, not to mention the countless kinds of fungi, protists, and prokaryotes. Even though in Darwin's day only a fraction of this number was known, the question why there are so many species and how they originated was already being asked. Lamarck had ignored the possibility of a multiplication of species in his *Philosophie Zoologique* (1809). For him, diversity was produced by differences in rates of adaptation, and new evolutionary lines originated by spontaneous generation, he thought. In Lyell's steady-state world, species number was constant, and new species were introduced to replace those that had become extinct. Any thought of the splitting of a species into several daughter species was absent among these earlier authors.

A solution to the problem of species diversification required an entirely new approach, and only the naturalists were in the position to find it. Leopold von Buch in the Canary Islands, Darwin in the Galapagos, Moritz Wagner in North Africa, and A. R. Wallace in Amazonia and the Malay Archipelago were the pioneers in this endeavor. They each discovered numerous populations that were in all conceivable intermediate stages of species formation. The sharp discontinuity between species that had so impressed John Ray, Carl Linnaeus, and others was now called into question by a continuity among species.

The problem of how these new species and incipient species come into being was clarified for Darwin by the Galapagos mockingbirds. These specimens showed that new species can originate by what we now call geographical (or allopatric) speciation. This theory of speciation says that new species may originate by the gradual genetic transformation of isolated populations. These isolated populations become in the course of time geographic races or subspecies, and Darwin realized that they may become new species, if isolated sufficiently long. By this thought Darwin founded a branch of evolutionism which, for short, we might designate as horizontal evolutionism, in contrast with the strictly vertical evolutionism of Lamarck. The two terms deal with two entirely different aspects of evolution, even though the processes responsible for these aspects proceed simultaneously. Vertical evolutionism deals with adaptive changes in the time dimension, while horizontal evolutionism deals with the origin of new diversity in the space dimension, that is, with the origin of incipient species and new species as populations move into new environmental niches. These species enrich the diversity of the organic world and are the potential founders of new higher taxa and of new evolutionary departures that will occupy new adaptive zones.

From 1837 on, when Darwin first recognized and solved the problem of the origin of diversity, this vertical/horizontal duality of the evolutionary process has been with us. Unfortunately, only a few authors have had the breadth of thought and experience of Darwin that would allow them to deal simultaneously with both aspects of evolution. Instead, paleontologists and geneticists have concentrated on vertical evolution, while the majority of naturalists have studied the origin of diversity as reflected in the horizontal process of the multiplication of species and the origin of higher taxa.

For Darwin, horizontal thinking about speciation permitted the solution of three important evolutionary problems: (1) why and how species multiply; (2) why there are discontinuities between major groups of organisms in nature, when the concept of gradual evolution would seem to imply countless subtle gradations between all groups; and (3) how higher taxa could evolve. But perhaps the most decisive consequence of the discovery of geographic speciation was that it led Darwin automatically to a branching concept of evolution. This is why branching entered Darwin's notebooks at such an early stage (see below).

Despite these insights of Darwin and the naturalists, the deep significance of the concept of speciation has been clarified only very slowly. As we will see in more detail in Chapter 3, in the decades after the publication of the *Origin of Species* Darwin himself vacillated in his understanding of how species multiply, and he continued to struggle with problems surrounding the origin of diversity for the rest of his life. Nevertheless, his realization that any evolutionary theory must somehow account for the multiplication of species was a pillar of Darwin's evolutionary thought from 1837 onward.

The Theory of Common Descent

The case of the species of Galapagos mockingbirds provided Darwin with an additional important new insight. The three species had clearly descended from a single ancestral species on the South American continent. From here it was only a small step to postulate that all mockingbirds were derived from a common ancestor—indeed, that every group of organisms descended from an ancestral species. This is Darwin's theory of common descent. By contrast, for those who accepted the concept of the *scala naturae* (scale of nature, or Great Chain of Being)—and in the eighteenth century this included most naturalists to a lesser or greater extent—all organisms were part of a single linear scale of ever-growing perfection. Lamarck still adhered, in principle, to this concept even though he allowed for some branching in his classification of the major phyla. Peter Simon Pallas and others had also published branching diagrams, but it required the categorical rejection of the *scala naturae* by Cuvier in the first and second decades of the nineteenth century before the need for a new way to represent organic diversity became crucial.

A group known as quinarians experimented with indicating relationship by osculating circles, but their diagrams did not fit reality at all well. The archetypes of Richard Owen and the *Naturphilosophen* strengthened the recognition of discrete groups in nature, but their use of the term "affinity" in relation to these groups remained meaningless prior to the acceptance of the theory of evolution. For Agassiz and Henri Milne-Edwards, branching reflected a divergence in ontogeny, so that the adult forms were far more different than the earlier embryonic stages. From all these examples it is evident that the static branching diagrams of nonevolutionists are no more indications of evolutionary thinking than branching flow charts in business or branching diagrams in administrative hierarchies.

Apparently very soon after Darwin understood that a single species of South American mockingbird had given rise to three daughter species in the Galapagos Islands, he seemed to realize that such a process of multiplication of species, combined with their continuing divergence, could in due time give rise to different genera and still higher categories. The members of a higher taxon then would be united by descent from a common ancestor. The best way to represent such a common descent would be a branching diagram. Already in the summer of 1837 Darwin clearly stated that "organized beings represent an irregularly branched tree" (*Notebook B:21*), and he drew several tree diagrams in which he even distinguished living and extinct species by different symbols.

By the time Darwin wrote the *Origin,* the theory of common descent had become the backbone of his evolutionary theory, not surprisingly so because it had extraordinary explanatory powers. Indeed, the manifestations of common descent as revealed by comparative anatomy, comparative embryology, systematics (in its search for "the natural system"), and biogeography became the main evidence for the occurrence of evolution in the years after 1859. Reciprocally, these biological disciplines, which up to 1859 had been primarily descriptive, now became causal sciences, with common descent providing an explanation for nearly everything that had previously been puzzling.

The concept of common descent was not entirely original with Darwin, however. Buffon had already considered it for close relatives such as horses and asses; but not accepting evolution, he had not extended this thought systematically. There are occasional suggestions of common descent in a number of other pre-Darwinian writers, but historians so far have not made a careful search for early adherents of common ancestry. The theory was definitely not upheld by Lamarck, who, although he proposed the occasional splitting of "masses" (higher taxa), never thought in terms of a splitting of species and regular branching. For him descent was linear descent within each phyletic line, and the concept of common descent was alien to him.

The theory of common descent, once proposed, is so simple and so obvious that it is hard to believe Darwin was the first to have adopted it consistently. Its importance was not only that it had such great explanatory powers but also that it provided a unity for the living world that had been previously missing. Up to 1859 people had been impressed primarily by the enormous diversity of life, from the lowest plants to the highest vertebrates, but this diversity took on an entirely different complexion when it was realized that it all could be traced back to a common origin. The final proof of this, of course, was not supplied until our time, when it was demonstrated that even bacteria have the same genetic code as animals and plants.

None of Darwin's theories was accepted as enthusiastically as common descent. Everything that had seemed to be arbitrary or chaotic in natural history up to that point now began to make sense. The archetypes of Owen and of the comparative anatomists could now be explained as the heritage from a common ancestor. The entire Linnaean hierarchy suddenly became quite logical, because it was now apparent that each higher taxon consisted of the descendants of a still more remote ancestor. Patterns of distribution that previously had seemed capricious could now be explained in terms of the dispersal of ancestors. Virtually all the proofs for evolution listed by Darwin in the *Origin* actually consist of evidence for common descent. To establish the line of descent of isolated or aberrant types became the most popular research program of the post-*Origin* period, and has largely remained the research program of comparative anatomists and paleontologists almost up to the present day. To shed light on common ancestors also became the program of comparative embryology. Even those who did not believe that ontogeny recapitulates phylogeny often discovered similarities in embryos that were obliterated in the adults. These similarities, such as the chorda in tunicates and vertebrates, or the gill arches in fishes and terrestrial tetrapods, had been totally mystifying until they were interpreted as vestiges of a common past. Though there are still a number of connections among higher taxa to be established, particularly among the phyla of plants and invertebrates, there is probably no biologist left today who would question that all organisms now found on the earth have descended from a single origin of life. This Darwin anticipated when he suggested that "all our plants and animals [have descended] from some one form, into which life was first breathed" (1859:484).

But perhaps the most important consequence of the theory of common descent was the change in the position of man. For theologians and philosophers alike, man was a creature apart from the rest of life. Aristotle, Descartes, and Kant agreed in this, no matter how much they disagreed in other aspects of their philosophies. Darwin, in the *Origin,* confined himself to the cautiously cryptic remark, "Light will be thrown on the origin of Man and his history" (p. 488). But Ernst Haeckel, T. H. Huxley, and in 1871 Darwin himself demonstrated conclusively that humans must have evolved from an ape-like ancestor, thus putting them right into the phylogenetic tree of the animal kingdom. This was the end of the traditional anthropocentrism of the Bible and of the philosophers. This application of the theory of common descent to humans, however, encountered vigorous opposition. To judge from contemporary cartoons, no Darwinian idea was less acceptable to the Victorians than the derivation of man from a primate ancestor. Yet today this derivation is not only remarkably well substantiated by the fossil record, but the biochemical and chromosomal similarity of man and the African apes is so great that it is puzzling why they are so relatively different in morphology and brain development. The primate origin of man, as first suggested by Darwin, immediately raised questions about the origin of mind and consciousness that are controversial to this day.

The Fate of Darwin's First Revolution

Darwin's theory of evolution as such (together with the nonconstancy of species), his theory of common descent, and his theory that species multiply were victorious within a remarkably short time. The victory of these three theories is the first Darwinian revolution. Within fifteen years of the publication of the *Origin* hardly a qualified biologist was left who had not become an evolutionist. The evolutionary origin of man was still unacceptable to some with religious commitments but was taken as firmly established by the anthropological profession.

Even though a theory of the multiplication of species is now taken for granted as an essential component of evolutionary theory, it is still controversial how this multiplication comes about. That much, if not most, speciation is geographical is generally accepted, but what other forms of speciation may also exist, and how frequent they are, is still an unsettled problem, as we will see in the next chapter.

CHAPTER THREE

How Species Originate

FROM THE SUMMER OF 1837 ON, Darwin collected notes on the great book he was going to write, and he referred to it in his notes and in his correspondence always as "the species book." When finally published in 1859, it was called by him *On the Origin of Species*. Darwin was fully conscious of the fact that the change from one species into another one was the most fundamental problem of evolution. Indeed, evolution was, almost by definition, a change from one species into another one. The belief in constant, unchangeable species was the fortress of antievolutionism to be stormed and destroyed. Once this was accomplished, it would not be difficult to adduce other evidence in favor of evolution.

In view of this central position of the problem of species and speciation in Darwin's life work, one would expect to find in the *Origin* a satisfactory and indeed authoritative treatment of the subject. This, curiously, one does not find. Indeed, the longer Darwin struggled with these concepts, the more he seemed to become confused. In the end, the *Origin* was a superb treatment of the theory of common descent and a great plea for the efficacy of natural selection, but it was vague and contradictory both on the nature of species and the mode of speciation.

What Is a Species?

After his crucial conversation with John Gould about the Galapagos mockingbirds in 1837, Darwin continued to struggle with the problem of how to define a species; but for that matter, so did virtually all other naturalists during the ensuing 150 years. It will help to understand Darwin's problem if I give a short overview of the four major species concepts that developed during this period.

THE TYPOLOGICAL SPECIES CONCEPT

A typological species is an entity that differs from other species by constant diagnostic characteristics. This was the species concept of Linnaeus and Lyell and was supported by those philosophers from Plato to modern times who consider species to be "natural kinds" or "classes." The concept was entirely consistent with the belief in creationism ("that which was separately created by God") and with the philosophy of essentialism ("that which has an essence of its own"; see Chapter 4). However, this species concept has three major practical weaknesses, which is the reason why it has been largely given up in recent times. First, it forces its adherents to consider as species even different variants within a population. Second, the invalidity of this concept is demonstrated by the high frequency of so-called sibling species—species which are indistinguishable on the basis of their appearance but which do not interbreed in nature. These species cannot be discriminated under the typological species concept. And third, it forces us to recognize as full species many local populations that differ by one diagnostic character from other populations of the species.

THE NOMINALIST SPECIES CONCEPT

According to this concept, only individual objects exist in nature. Similar objects or organisms are bracketed together by a name, and this subjective action of the classifier determines which objects are combined into one species. Species, therefore, are merely arbitrary mental constructs. That species have no reality in nature has been the claim of the nominalists from the Middle Ages right up to the writings of some recent philosophers. By contrast, the reality of species has been supported consistently by naturalists up to the present day. There is no more devastating refutation of the nominalistic claim than the fact that primitive natives in New Guinea, with a Stone Age culture, recognize as species exactly the same entities of nature as Western taxonomists. If species were something purely arbitrary, it would be totally improbable for representatives of two drastically different cultures to arrive at identical species delimitations.

THE EVOLUTIONARY SPECIES CONCEPT

Paleontologists who study species distributed in the time dimension have always been rather dissatisfied with a species definition derived from the nondimensional species concept of the local naturalist. What paleontologists were looking for was a species concept that would be particularly suitable for the discrimination of fossil species. This need eventually led the naturalist G. G. Simpson (1961:153) to this species definition: "An evolutionary species is a lineage (an ancestral–descendent sequence of populations) evolving separately from others and with its own unitary evolutionary role and tendencies." Wiley (1980, 1981) and Willmann (1985) have proposed slightly modified versions of Simpson's definition.

An evolutionary definition of species has not been widely accepted because it is only applicable to monotypic species, since all isolates that "evolve separately" would also have to be recognized as species. Furthermore, how is one to describe and determine the "unitary evolutionary role and tendencies" of a population or taxon? The main objective of the evolutionary species definition was to permit a clear delimitation of a species in the time dimension, but this hope turned out to be illusory in all cases of gradual species transformation. The exact determination of the origin of a new species can be made only in cases of instantaneous speciation (as in polyploidy), and the exact endpoint of a species can only be determined in the case of extinction. The biological species definition, as we will see, is equally qualified to determine these two points, and has other advantages going for it as well.

THE BIOLOGICAL SPECIES CONCEPT

This species concept is based on the observation of local naturalists that at a given locality, populations of different species coexist but do not interbreed with each other. This I articulated in the definition: "Species are groups of interbreeding natural populations that are reproductively isolated from other such groups." Statements in Darwin's notebooks show that by 1837 he had given up the typological species concept and had developed a species concept based on reproductive isolation. For instance, "My definition of species has nothing to do with hybridity, is simply an instinctive impulse to keep separate, which will no doubt be overcome [or else no hybrids would ever be produced], but until it is, these animals are distinct species" (C:161). There are repeated references in the notebooks to mutual "repugnance" of species to intercrossing. "The dislike of two species to each other is evidently an instinct; and this prevents breeding" (B:197). "Definition of species: one that remains at large with constant characters, together with other beings of very near structure" (B:213). For Darwin, species status had little if anything to do with degree of difference: "Hence species may be good ones and differ scarcely in any external character" (B:213).

Darwin adhered to this biological species concept for about the next fifteen years. But then he became rather confused, particularly after he tried to apply his zoological findings to plants. As we shall see, he considered the *variety* (which in animals is a subspecies) to be an incipient species, and he encountered no difficulties with this concept as long as he dealt with animals. However, when through Hooker and other botanical friends Darwin came to study varieties of plants, he did not realize that the botanical terminology was rather different from that of zoology. Plant varieties very often were individual variants within a local population, and to consider them incipient species not only caused problems for the explanation of speciation but also for the delimitation of species against varieties, and of species against one another. There were a number of other developments in these years which induced Darwin to give up his biological species concept and return to a more or less typological species definition (Sulloway 1979). The twenty-five pages on species and speciation in his unfinished big book manuscript (Darwin 1975:250–274) contain so many contradictions that they are almost embarrassing to read.

The species concept at which Darwin finally arrived is clearly described in the *Origin*. There is nothing left of the biological criteria of the notebooks, and his characterization of the species now is a mixture of the typological and nominalist species definitions.

> Systematists . . . will not be incessantly haunted by the shadowy doubt whether this or that form be in essence a species. This I feel sure . . . will be no slight relief. . . . Systematists will have only to decide . . . whether any form be sufficiently constant and distinct from other forms, to be capable of definition; and, if definable, whether the differences be sufficiently important to deserve a specific name . . . The only distinction between species and well-marked varieties is, that the latter are known, or believed, to be connected at the present day by intermediate gradations . . . In short, we shall have to take species in the same manner as those naturalists treat genera, who admit that genera are merely artificial combinations made for convenience. This may not be a cheering prospect; but we shall at least be freed from the vain search for the undiscovered and undiscoverable essence of the term species" (1859:484–485).

The example set by Darwin was followed by just about every taxonomist and evolutionist in the nineteenth century except for a few enlightened field naturalists. It clearly was the species concept of the Mendelians. The counter movement in the twentieth century, initiated by Jordan, Poulton, Stresemann, and other progressive workers, ultimately resulted in the biological species concept's being widely accepted.

However, each of the three major species concepts (typological, evolutionary, and biological) has a certain legitimacy in some areas of biological research, even today. Which of them one adopts may depend on the type of research one does. The museum taxonomist, as well as the stratigrapher, may find the typological species concept most useful, never mind how clearly it is refuted by the existence of sibling species and strikingly different variants (phena). But anyone working with living populations, restricted to one place and time, finds any species concept other than the biological one to be unsatisfactory. Finally, the paleontologist, part of whose endeavor it is to delimit fossil species taxa in the vertical sequence of strata, cannot help but pay attention to the time dimension.

Modern biologists are almost unanimously agreed that there are real discontinuities in organic nature, which delimit natural entities that are designated as species. Therefore, the species is one of the basic foundations of almost all biological disciplines. Each species has different biological characteristics, and the analysis and comparison of these differences is a prerequisite for all other research in ecology, behavioral biology, comparative morphology and physiology, molecular biology, and indeed all branches of biology. Whether he realizes it or not, every biologist works with species.

What Is Speciation?

Paleontologists, conditioned to vertical thinking, considered speciation to be the change of a phyletic lineage over time. But since there is a steady extinction of species, where do the new species come from? This has been the problem from Lamarck and Lyell on. The answer which Darwin found was that species not only evolve in time but also multiply. Speciation, since Darwin, has come to mean the production of new, reproductively isolated populations.

That the multiplication of species was a major evolutionary phenomenon became clear to Darwin as soon as he followed up the consequences of his new understanding of the taxonomy of the Galapagos mockingbirds. But it was really not until the coming of the new systematics and the evolutionary synthesis (see Chapter 9) that speciation became the focal point of evolutionary research at the species level.

Considering Darwin's uncertainties about the nature of species, it is perhaps not surprising that he vacillated about the mode of speciation. It must be remembered that at first Darwin believed in Lyell's sudden, saltational "introduction" of new species. It was John Gould's pronouncement in March 1837 that each of the three populations of mockingbirds in the Galapagos Islands was a separate species that started Darwin along an entirely new path of thinking about the origin of diversity. Obviously, the different mockingbirds on the three islands in the Galapagos had all descended from colonists of the South American mainland species; but the three populations had evolved in a slightly different manner on each of the three islands. This led Darwin to adopt the theory of speciation by the gradual modification of populations that were geographically isolated from their parent species. And this was Darwin's theory of speciation until the early 1850s. It was not a new theory, because a similar one had previously been proposed by von Buch (1825). Later, Wagner (1841) and Wallace (1855) also independently arrived at the same conclusion.

Darwin considered isolation on islands as the principal speciation mechanism, and he seems to have had difficulties in explaining speciation on continents. At one time, to account for the rich species diversity in South Africa, he postulated large-scale geological changes—up and down movements of the crust—during which South Africa was temporarily converted into an archipelago, setting the stage for abundant geographical speciation.

Just as Darwin's botanical researches induced him to change his species concept, so did these researches, at least in part, induce Darwin to accept, in addition to geographic speciation, a process we now call sympatric speciation. The exact reasoning by which Darwin arrived at this new explanation is still not fully explored (Sulloway 1979; Mayr 1982; Kohn 1985). Owing to the prevailing typological thinking, geographic races of animals were called varieties in Darwin's time. In plants, however, as mentioned above, the term variety had a dual meaning. It was applied not only to geographic races (subspecies) but also to individual variants (phena) within a single population. Darwin, on the basis of his zoological studies, had concluded that varieties of animals were incipient species. When he transferred this concept of the role of varieties in speciation to plants, he was induced to consider also individual variants within a population as incipient species. Darwin concluded, therefore, that in addition to geographic speciation there is also a process of sympatric speciation, which he defined as the origin of a new species by ecological specialization within the range of the parental species. The survival and flourishing of the new sympatric species, it was said, was made possible by its shift into a new niche, thus removing it from competition with the parental species. Natural selection would lead to an ever greater difference (character divergence) of the two competing new species. Darwin developed this concept of sympatric speciation by character divergence in the years 1854 to 1858 (Kohn 1985) and continued to support it even though once in a while he would concede that maybe geographic isolation was nearly always necessary.

It was this claim of sympatric speciation that involved Darwin in a bitter controversy with the explorer and naturalist Moritz Wagner (1868, 1889), who insisted that geographic isolation was absolutely indispensible for speciation. Unfortunately Wagner beclouded the issue by also insisting that natural selection could not operate unless the population was isolated. What we now know about speciation indicates that Wagner's arguments on the whole were more valid than Darwin's. Nevertheless, Darwin's insistence on the frequency of sympatric speciation prevailed, and prior to the evolutionary synthesis sympatric speciation was considered a frequent and, as far as the entomological literature is concerned, perhaps the prevailing mode of speciation.

In his early notebook statements on species (B:197, 213; C:161), Darwin clearly saw species as reproductively isolated entities. By the 1850s, however, his attention concentrated on species as adapted entities. He would describe how they became adapted to a new niche, but he failed to account for their reproductive isolation from the parental species. His reasoning was aggravated by a return from populational to typological thinking (see Chapter 4). "If a variety were to flourish so as to exceed in number the parent species, it would then rank as the species and the species as the variety, or it might come to supplant and exterminate the parent species, or both might coexist, and both rank as independent species" (1859:52). How could the new variety coexist as an independent species unless it had acquired reproductive isolation? This question is particularly troublesome, given that Darwin, in an argument about the origin of isolating mechanisms among species, definitely rejected the capacity of natural selection to establish a reproductive barrier (Mayr 1959b; Sulloway 1979).

Looking over all of Darwin's writings on species and speciation, one cannot escape the impression that he was rather confused about the issues and that he not infrequently contradicted himself. A major reason for his confusion can be traced to his puzzlement over the origin of genetic variation. The clear statements about mechanisms of speciation that we now can make more than 125 years later are based on our understanding (as far as it goes) of genetics. Despite his failure to reach a definitive conclusion about species and speciation, Darwin deserves credit for having pointed out the problems and for having clearly stated various alternative solutions.

CHAPTER FOUR

Ideological Opposition to Darwin's Five Theories

IN BOTH SCHOLARLY and popular literature one frequently finds references to "Darwin's theory of evolution," as though it were a unitary entity. In reality, Darwin's "theory" of evolution was a whole bundle of theories, and it is impossible to discuss Darwin's evolutionary thought constructively if one does not distinguish its various components. The current literature can easily leave one perplexed over the disagreements and outright contradictions among Darwin specialists, until one realizes that to a large extent these differences of opinion are due to a failure of some of these students of Darwin to appreciate the complexity of his paradigm.

The term "Darwinism," as we will see in Chapter 7, has numerous meanings depending on who has used the term and at what period. A better understanding of the meaning of this term is only one reason to call attention to the composite nature of Darwin's evolutionary thought. Another reason is that one cannot answer the question correctly of how and when "Darwinism" was accepted in different countries throughout the world unless one focuses on the various Darwinian theories separately. What Darwin presented in 1859 in the *Origin* was a compound theory, whose five subtheories had very different fates in the eighty years after Darwin.

One particularly cogent reason why Darwinism cannot be a single monolithic theory is that organic evolution consists of two essentially independent processes, as we have seen: transformation in time and diversification in ecological and geographical

space. The two processes require a minimum of two entirely independent and very different theories. That writers on Darwin
have nevertheless almost invariably spoken of the combination of
these various theories as "Darwin's theory" in the singular is in
part Darwin's own doing. He not only referred to the theory of
evolution by common descent as "my theory," but he also called
the theory of evolution by natural selection "my theory," as if
common descent and natural selection were a single theory.

Discrimination among his various theories has not been helped
by the fact that Darwin treated speciation under natural selection
in chapter 4 of the *Origin* and that he ascribed many phenomena,
particularly those of geographic distribution, to natural selection
when they were really the consequences of common descent.
Under the circumstances I consider it necessary to dissect Darwin's conceptual framework of evolution into a number of major
theories that formed the basis of his evolutionary thinking. For
the sake of convenience I have partitioned Darwin's evolutionary
paradigm into five theories, but of course others might prefer a
different division. The selected theories are by no means all of
Darwin's evolutionary theories; others were, for instance, sexual
selection, pangenesis, effect of use and disuse, and character divergence. However, when later authors referred to Darwin's
theory they invariably had a combination of some of the following five theories in mind:

(1) *Evolution as such.* This is the theory that the world is not constant
    nor recently created nor perpetually cycling but rather is steadily
    changing and that organisms are transformed in time.
(2) *Common descent.* This is the theory that every group of organisms
    descended from a common ancestor and that all groups of orga-
    nisms, including animals, plants, and microorganisms, ulti-
    mately go back to a single origin of life on earth.
(3) *Multiplication of species.* This theory explains the origin of the
    enormous organic diversity. It postulates that species multiply,
    either by splitting into daughter species or by "budding," that is,
    by the establishment of geographically isolated founder popula-
    tions that evolve into new species.

(4) *Gradualism*. According to this theory, evolutionary change takes place through the gradual change of populations and not by the sudden (saltational) production of new individuals that represent a new type.

(5) *Natural selection*. According to this theory, evolutionary change comes about through the abundant production of genetic variation in every generation. The relatively few individuals who survive, owing to a particularly well-adapted combination of inheritable characters, give rise to the next generation.

For Darwin himself these five theories were apparently a unity, and someone might claim that indeed these five theories are a logically inseparable package and that Darwin was quite correct in treating them as such. This claim, however, is refuted by the fact, as demonstrated in Table 1, that most evolutionists in the immediate post-1859 period—that is, authors who had accepted the first theory—rejected one or several of Darwin's other four theories. This shows that the five theories are not one indivisible whole.

There are several reasons why Darwin's five theories had such different fates. One reason is that abundant evidence was already available to support some of them. This was true, for example,

TABLE I

The composition of the evolutionary theories of various evolutionists. All these authors accepted a fifth component, that of evolution as opposed to a constant, unchanging world.

	Common descent	Multiplication of species	Gradu-alism	Natural selection
Lamarck	No	No	Yes	No
Darwin	Yes	Yes	Yes	Yes
Haeckel	Yes	?	Yes	In part
Neo-Lamarckians	Yes	Yes	Yes	No
T. H. Huxley	Yes	No	No	(No)[a]
de Vries	Yes	No	No	No
T. H. Morgan	Yes	No	(No)[a]	Unimportant

a. Parentheses indicate ambivalence or contradiction.

for the theory of common descent. In the case of the theory of natural selection, however, the evidence for or against was contradictory, and it required generations of research time before the controversial points could be settled. A more important reason, however, for the delay in the acceptance of some of Darwin's theories was that they were in conflict with certain ideologies dominant at Darwin's time.

During the three centuries preceding the publication of the *Origin,* Europe had been in the throes of a continuous intellectual upheaval, punctuated by the Scientific Revolution of the sixteenth and seventeenth centuries and the Enlightenment of the eighteenth century. Why did it take so long for evolution to be seriously proposed? And why did Darwinism face such an uphill battle after it was proposed? The reason is that Darwin challenged some of the basic beliefs of his age. Four of them were pillars of Christian dogma:

(1) *A belief in a constant world.* In spite of Lamarck and the *Naturphilosophen,* it was still widely, if not almost universally, accepted in 1859 that except for minor perturbations (floods, volcanism, mountain building) the world had not changed materially since creation. And in spite of Buffon, Kant, Hutton, Lyell, and the ice age theory, the prevailing opinion was still that the world had been created rather recently.

(2) *A belief in a created world.* Species and other taxa were believed to be unchanging, and therefore the existing diversity of the living world could only be due to an act of creation. This was a single creation as believed by the orthodox Christians or repeated creations, either of whole biota as believed by the so-called progressionists (for example, Agassiz), or of individual species as proposed by Lyell.

(3) *A belief in a world designed by a wise and benign Creator.* Even though the world had its imperfections, it was the best of the *possible* worlds, according to Leibniz. The adaptation of organisms to their physical and living environment was perfect because it had been designed by an omnipotent Creator.

(4) *A belief in the unique position of man in the creation.* The world was anthropocentric in the eyes of the Christian religion as well as in

the view of the foremost philosophers. Man had a soul, something animals did not have. There was no possible transition from animal to man.

In addition to these four religious beliefs, there were three secular philosophies that were also in conflict with certain of Darwin's theories. These secular beliefs were:

(5) *A belief in the philosophy of essentialism* (see below).

(6) *A belief in an interpretation of the causal processes of nature as they had been elaborated by the physicists* (see Chapter 5).

(7) *A belief in "final causes" or teleology* (see Chapter 5).

The theories of evolution proposed by Darwin challenged all seven of these traditional and well-entrenched views, though not every one of Darwin's five theories was in conflict with all of them.

External Factors in Scientific Revolutions

For the last sixty years there has been a good deal of controversy in the history of science over the question whether scientific revolutions or, indeed, any conceptual changes in science are due to discoveries and new observations made within the field ("internalism") or are the result of outside influences ("externalism"). The Marxist thesis, first promoted by Hessen (1931), that the socioeconomic milieu is decisive in initiating conceptual changes in science was particularly actively debated during part of this period.

Researches during this period have made it clear that one must distinguish between two sets of external factors: socioeconomic and ideological ones. It has also become quite clear that these two sets of factors differ fundamentally in their impact on science. Socioeconomic factors, it would seem, have very little influence on the acceptance or rejection of scientific ideas, as was shown by Mayr (1982:4–5) and by many other historians. If the industrial revolution and the socioeconomic situation had been responsible for the theory of natural selection, this theory should have been most palatable to the British public of the mid-nineteenth century. But this was not the case at all. On the contrary, the theory of natural selection, as we shall see, was almost unanimously rejected by Darwin's contemporaries. It was evidently not a reflection of the socioeconomic situation.

The other set of external factors—ideological ones—by contrast, had a powerful effect on the acceptance of Darwin's theories. What is remarkable and rarely sufficiently emphasized is that it was not the prevailing ideologies almost universally adopted in Britain (and much of the rest of the world) in the first half of the nineteenth century that inspired Darwin and gave him his new theories. On the contrary, the prevailing *Zeitgeist* was utterly opposed to Darwin's thought and prevented a universal acceptance of some of his new ideas for more than a hundred years. Indeed, as shown by the many recently published books and papers that still attempt to refute Darwinism, many Darwinian ideas are still not yet full accepted, owing to the continuing power of the opposing ideologies.

From Essentialism to Population Thinking

Of the seven ideologies challenged by Darwin's theories, none was more deeply entrenched than the philosophy of essentialism. Essentialism had dominated Western thinking for more than 2000 years, going back to the geometric thinking of the Pythagoreans. They pointed out that a triangle, regardless of the combination of angles, always has the form of a triangle. It is discontinuously different from a quadrangle or any other kind of polygon. The triangle is one of the limited number of possible forms of a polygon. In an analogous manner, all the variable phenomena of nature, according to this thinking, are a reflection of a limited number of constant and sharply delimited *eide* or essences. Essentialism, as a definite philosophy, is usually credited to Plato, even though he was not as dogmatic about it as some of his later followers, for instance the Thomists. Plato's cave allegory of the world is well known: What we see of the phenomena of the world corresponds to the shadows of the real objects cast on the cave wall by a fire. We can never see the real essences. Variation is the manifestation of imperfect reflections of the underlying constant essences.

All of Darwin's teachers and friends were, more or less, essentialists. For Lyell, all nature consisted of constant types, each created at a definite time. "There are fixed limits beyond which the descendants from common parents can never deviate from a certain type." And he added emphatically: "It is idle . . . to dispute about the abstract possibility of the conversion of one species into another, when there are known causes, so much more active in their nature, which must always intervene and prevent the actual accomplishment of such conversions" (Lyell 1835:162). For an essentialist there can be no evolution: there can only be a sudden origin of a new essence by a major mutation or saltation.

Virtually all philosophers up to Darwin's time were essentialists. Whether they were realists or idealists, materialists or nominalists, they all saw species of organisms with the eyes of an essentialist. They considered species as "natural kinds," defined by constant characteristics and sharply separated from one another by bridgeless gaps. The essentialist philosopher William Whewell stated categorically, "Species have a real existence in nature, and a transition from one to another does not exist" (1840, 3:626). For John Stuart Mill, species of organisms are natural kinds, just as inanimate objects are, and "kinds are classes between which there is an impassable barrier."

Essentialism's influence was great in part because its principle is anchored in our language, in our use of a single noun in the singular to designate highly variable phenomena of our environment, such as mountain, home, water, horse, or honesty. Even though there is great variety in kinds of mountain and kinds of home, and even though the kinds do not stand in direct relation to one another (as do the members of a species), the simple noun defines the class of objects. Essentialistic thinking has been highly successful, indeed absolutely necessary, in mathematics, physics, and logic. The observation of nature seemed to give powerful support to the essentialists' claims. Wherever one looked, one saw discontinuities—between species, between genera, between orders and all higher taxa. Such gaps as between birds and mammals, or beetles and butterflies, were mentioned often by Darwin's critics.

In daily life we largely proceed essentialistically (typologically) and become aware of variation only when we compare individuals. He who speaks of "the Prussian," "the Jew," "the intellectual" reveals essentialistic thinking. Such language ignores the fact that every human is unique; no other individual is identical to him.

It was Darwin's genius to see that this uniqueness of each individual is not limited to the human species but is equally true for every sexually reproducing species of animal and plant. Indeed, the discovery of the importance of the individual became the cornerstone of Darwin's theory of natural selection. It eventually resulted in the replacement of essentialism by population thinking, which emphasized the uniqueness of the individual and the critical role of individuality in evolution. Darwin no longer asked, as had Agassiz, Lyell, and the philosophers, "What is good for the species?" but "What is good for the individual?" (Ghiselin 1969). And variation, which had been irrelevant and accidental for the essentialist, now became one of the crucial phenomena of living nature.

From Saltationism to Gradualism

Many of Darwin's contemporaries who accepted the fact of evolution nevertheless were incapable of population thinking, owing to their ideological commitment to essentialism. As we have seen, they accepted instead a concept of evolution based on the sudden production of new species through saltations. Saltational evolution is a necessary consequence of essentialism: if one believes in evolution and in constant types, only the sudden production of a new type can lead to evolutionary change. That such saltations can occur and indeed that their occurrence is a necessity

are old beliefs. Almost all theories of evolution described by Osborn in his history of evolution, *From the Greeks to Darwin* (1894),
were saltational theories, that is, theories of the sudden origin of
new kinds. After the publication of the *Origin,* many biologists
who accepted evolution as such (Darwin's first theory) nevertheless, because they were essentialists, turned to saltational theories
to explain the process of evolution.

One can distinguish three kinds of saltationist theories: (1) extinct species are replaced by newly created ones that are more or
less at the same level as those that they replace (Lyell 1830–1833);
(2) extinct species are replaced by new creations at a higher level
of organization (progressionists, such as William Buckland,
Sedgwick, Hugh Miller, Agassiz); (3) new species originate
through saltations of pre-existing species (E. Geoffroy Saint-
Hilaire, Darwin in Patagonia, Galton, Goldschmidt).

Saltational evolution is best called "transmutation," because the
production of new species or new types is discontinuous, owing
to the sudden creation of a new essence. But transmutation
should not be confused with another concept of evolution known
as transformational evolution. According to this concept, evolution consists of the gradual transformation of a thing from one
condition to another. Lamarck's concept of evolution was transformational, designating a completely gradual process, a change
due to a trend toward perfection or adjustment to the environment. To the best of my knowledge Lamarck was the first author
to propose a consistent theory of gradual transformation. After
1800, but before 1859, the idea of gradual evolution was accepted
by a considerable number of authors on the Continent, but in a
vague manner and unsupported by adequate evidence.

As Lewontin (1983) has pointed out, Darwin, by contrast, introduced a new concept of evolution that was entirely different
from saltational evolution or transformational evolution. According to Darwin's concept, which we can designate as "variational evolution," variations are produced in every generation,
and evolution takes place because only a small number of variants
survive to reproduce. No longer is a concrete object transformed,

as in transformational evolution, but a new beginning is made in every generation. Indeed, evolution is a two-step phenomenon, the first step in each generation being responsible for the production of variation, which is then sorted in the second step, selection proper (see Chapter 6). Thus, strictly speaking, Darwinian evolution is discontinuous because a new start is made in every generation when a new set of individuals is produced. That evolution nevertheless appears to be totally gradual is because it is populational and depends on sexual reproduction among the members of the population. Such evolution is not necessarily progressive; it is an opportunistic response to the moment; hence, it is unpredictable.

Darwin's Growing Commitment to Gradualism

The concept of the gradual transformation of a population was not entirely new with Darwin. The occurrence of continuity had been stressed by some authors as far back as Aristotle, with his principle of plenitude (Lovejoy 1936). It was reflected in the concept of the *scala naturae,* and even such an arch-essentialist as Linnaeus stated that the orders of plants were touching each other like countries on a map. Lamarck was the first person to apply the principle of gradualism to the origin of the hierarchy of life, but there is no evidence that Darwin derived his gradualistic thinking from Lamarck.

Then how did Darwin arrive at the concept of gradual evolution? References to gradual changes are scattered through Darwin's notebooks from early on (Kohn 1980). For instance, Darwin considered the changes of organisms either to be produced directly by the environment or to be at least an answer to the changes in the environment. Hence, "The changes in species must be very slow, owing to physical changes slow" (*Notebook C:*17). Gradualness was also favored by Darwin's conclusion that changes in habit or behavior may precede changes in structure (*Notebook C:*57, 199). At that time Darwin still believed in a principle called Yarrell's Law (named after William Yarrell, a naturalist and animal breeder), according to which it takes many generations for the effects of the environment or of use and disuse to become strongly hereditary. As Darwin stated, "Variety when long in blood, gets stronger and stronger" (*Notebook C:*136). Various other sources for Darwin's gradualist thinking have been suggested in the recent literature, such as the writings of J. B. Sumner (Gruber 1974:125), or Leibniz's principle of plenitude (Stanley 1981). But to me it seems more likely that Darwin arrived at his gradualism owing to two major influences. One was Lyell's uniformitarianism, which Darwin extended from geology to the organic world. The other influence consisted of his own empirical researches.

At least three observations may have been influential: (1) the slightness of the differences among the mockingbird populations on the three Galapagos islands and the South American mainland, as well as a similarly slight difference among many varieties and species of animals; (2) the barnacle researches, where Darwin complained incessantly about the extent to which species and varieties were intergrading; (3) Darwin's work with races of domestic pigeons, where he convinced himself that even the most extreme races (which, if found in nature, would be unhesitatingly placed by taxonomists in different genera) were nevertheless the product of painstaking, long-continued, gradual, artificial selection. In his *Essay* of 1844 Darwin argues in favor of gradual evolution by analogy with what is found in domesticated animals and plants. And he postulates that "there must have existed intermediate forms between all the species of the same group, not differing more than recognized varieties differ" (p. 157).

Finally, Darwin had didactic reasons for insisting on the slow accumulation of rather small steps. He answered the argument of his opponents that one should be able to "observe" evolutionary change owing to natural selection by saying: "As natural selection acts solely by accumulating slight successive favorable variations, it can produce no great or sudden modifications; it can act only by very short and slow steps" (1859:471).

Thus, Darwin's rejection of essentialism and the general emergence of population thinking strengthened his adherence to gradualism and led him to totally reject saltations. As soon as one adopts the concept that species evolve as populations are transformed, owing to the differential reproductive success of unique individuals over generations—and this is what Darwin increasingly believed—one is automatically forced also to believe that evolution must be gradual. Gradualism and population thinking probably were originally independent strands in Darwin's conceptual framework, but eventually they reinforced each other powerfully, just as essentialism and saltationism reinforced one another in the thinking of many of Darwin's opponents.

Darwin's totally gradualist theory of evolution—that not only species but also higher taxa arise through gradual transformation—immediately encountered strong opposition. Even some of Darwin's closest friends were unhappy about it. T. H. Huxley wrote to Darwin on the day before the publication of the *Origin:* "You have loaded yourself with an unnecessary difficulty in adopting *Natura non facit saltum* [Nature makes no jumps] so unreservedly" (Huxley 1900 2:27). In spite of the urgings of Huxley, Galton, Kölliker, and other contemporaries, Darwin insisted on the gradualness of evolution, even though he was fully aware of the controversial nature of this concept. Furthermore, Darwin's adherence to gradualism became stronger with time; eventually (after the 1867 critique by F. Jenkin) he minimized even more the evolutionary role of drastic variations ("sports").

But essentialism and saltationism continued to be widely adopted. After Darwin's death the concept of gradualism became even less popular than it had been in Darwin's own time. This began with William Bateson's 1894 book and reached a climax with the mutationist theories of the Mendelians (see Chapter 9). Both Bateson and de Vries missed no opportunity to make fun of Darwin's belief in gradual evolution and upheld instead evolution by macromutations (Mayr and Provine 1980). A mild popularity of saltationist theories continued right through the evolutionary synthesis (Goldschmidt 1940; Willis 1940; Schindewolf 1950). The naturalists were the main supporters of gradual evolution, which they encountered everywhere in the form of geographic variation. Eventually, geneticists arrived at the same conclusion through the discovery of ever slighter mutations.

Defining gradualism as populational evolution—and this is what Darwin basically had in mind—permits us to say that in spite of all the opposition to him, Darwin ultimately prevailed even with his fourth evolutionary theory. The only exceptions to gradualism that are clearly established are cases of stabilized hybrids that can reproduce without crossing.

Nothing is said in the theory of gradualism about the precise rate at which the change may occur. Darwin was aware of the fact that evolution could sometimes progress rapidly, but, as Andrew Huxley (1981) has pointed out, evolution could also contain periods of complete stasis "during which these same species remained without undergoing any change." (In a well-known diagram in the *Origin*, Darwin lets one species (F) continue unchanged through 14,000 generations or even through a whole series of geological strata. Understanding the independence of gradualness and evolutionary rate is important for evaluating the theory of punctuated equilibria put forth in 1972 by Stephen J. Gould and Niles Eldredge (see Chapter 10).

CHAPTER FIVE

The Struggle against Physicists and Philosophers

E SSENTIALISM WAS NOT the only ideology Darwin had to overcome. A concept of science had developed since the seventeenth century that was completely dominated by physics and mathematics. The philosophers from Bacon and Descartes to Locke and Kant entirely agreed with the physical scientists from Galileo and Newton to Lavoisier and Laplace that the ideal of science should be to establish mathematically formed theories that were based on universal laws. The possibility of proof and of exact prediction were the tests for the goodness of a scientific explanation. Newton's physics was the shining example of good science. Every British scientist and philosopher, at the time when Darwin was developing his ideas on evolution, agreed in this concept of science, and there are indications that Darwin did his best, particularly after reading and rereading John Herschel's *Discourse,* to live up to his ideal (Ruse 1979).

In spite of all these efforts, Darwin's empirical researches led to results that were in conflict with most of the basic assumptions of physicalism. The physicalists were essentialists, a philosophy that Darwin totally rejected. Instead, Darwin developed population thinking, a mode of thinking utterly alien to the physicalists. The physicists at that time were strict determinists; prediction was not only possible but was the very test of the validity of theories. Evolutionary processes, by contrast, involved a considerable chance element: they were probabilistic, and hence they did not permit absolute prediction. All evolutionary phenomena could be explained only by inferring past historical events, a consideration absent (at that period) in the physical sciences. The probabilistic nature of the evolutionary process was so alien to the thinking of the physicists that Herschel referred to natural selection as the theory of the "higgledy-piggledy" (F. Darwin 1888, 2:240).

Darwin's findings completely undermined the physicalist concept of laws. The order and harmony of the created universe made the physical scientists search for laws, for wise regulations in the running of the universe installed by the Creator. To serve his Creator best a physicist studied His laws and their working. In this tradition Darwin refers in the *Origin of Species* no fewer than 106 times in 490 pages to laws controlling certain biological processes. But Darwin's "laws" were not the laws of the deists but were either simple facts or regular processes. No longer relying on universal laws, Darwin had no problem in accepting statistical generalizations. It was a complete rejection of Cartesian-Newtonian determinism.

The role Darwin assigned to chance has never been properly analyzed. The deterministic spirit of science at his time was in complete conflict with Darwin's findings, which showed how strong a role in evolution was played by chance. In the case of variations, he distinguished between those being accidental as far as their "purpose" or selective value is concerned, and others that are "accidental as to their cause of origin" (F. Darwin 1888 I:314). A similar train of thought is expressed in the *Variation of Animals and Plants* (1868 2:431). It is evident that Darwin accepted the strict working of what he called natural laws at the physiological level but was aware of chance (stochastic) processes at the organismic level.

The realization that neither essentialism nor determinism nor any other aspect of physicalism was a valid ideology was of the utmost importance for the further development of Darwin's thought. Darwin could have never adopted natural selection as a major theory, even after he had arrived at the principle on a largely empirical basis, if he had not rejected essentialism and physicalism. But there was one other impeding ideology Darwin had to refute in order to be able to adopt natural selection, and that was the finalistic or teleological worldview.

Final Causes

Again and again the statement has been made that "Darwin was no philosopher," even by an otherwise so perceptive author as G. G. Simpson (1964:50). In fact, Darwin was keenly interested in philosophy and, as we have seen, attempted to follow in his own writings the best advice of the philosophers of science of his day. Admittedly, he never published an essay or volume explicitly devoted to an exposition of his philosophical ideas, but in his scientific works he systematically demolished one after the other of the basic philosophical concepts of his time and replaced them with revolutionary new concepts.

In retrospect it is rather surprising to what an extent the contemporary philosophers who were involved with scientific ideas ignored Darwin or at least failed either to incorporate his concepts into their own systems or at least to try seriously to refute him adequately. This is as true for the British philosophers Herschel, Whewell, Mill, and Jevons, as well as for the various German philosophers and philosophical biologists, the followers of Kant, the later *Naturphilosophen,* the mechanoteleologists (Lenoir 1982), and the mechanists (DuBois-Reymond, Helmholtz, Sachs, Ludwig, Loeb). To a large extent these philosophers failed to recognize the importance of Darwin's ideas because of their commitment to the philosophies of essentialism and finalism.

From the days of the earliest philosophers it was widely believed that the world must have a purpose because, as Aristotle had said, "Nature does nothing in vain," and neither, a Christian would say, does God. Any change in this world, they would say, is due to final causes that move the particular object or phenomenon toward an ultimate goal. The development of an organism from the fertilized egg to the adult stage was frequently cited, from Aristotle on, as an illustration of this striving toward a goal.

It was almost universally believed that everything in nature, particularly all directional processes, moved in an analogous manner toward a predetermined end. Those who adhered to this view have been designated teleologists, or "finalists."

The thinkers of the Scientific Revolution in the sixteenth and seventeenth centuries were fascinated by motion. They worked out the laws of falling bodies and the motion of the planets around the sun; for them, the world was a world of motion controlled by eternal laws. God had instituted these laws at the time of creation, but from that point on laws kept the world moving. God was the final cause of everything, but He ruled the world through His laws and not by continuous intervention. René Descartes was one of the chief spokesmen for this strictly physicalist or mechanistic worldview, which was primarily that of the physicists. It was, however, more or less adopted even by the naturalist Buffon and carried to the most extreme consequences by Baron Holbach. Even those who adopted this concept of a mechanized world had certain misgivings about applying it to the living world. Buffon, for instance, was fully aware of the conflict between the mechanized world picture and many phenomena encountered in the study of organisms. Yet any alternative view was unacceptable to him.

Those who were dissatisfied with a strictly mechanistic world, entirely run by laws, developed a different explanatory framework. They credited God with a much larger role in designing the world down to the last detail and in effecting the changes that had taken place since creation. It was distasteful to them to remove God from the running of His world and to replace Him by the efficient causes of His laws. Not only that, but they also found it inconceivable that the observed harmony of nature and all the mutual adaptations of organisms to one another could be due simply to efficient causes. Their answer was to stress the elaborateness of the original design of the world to a far greater degree than had been done by the mechanists. No matter where you looked in nature, they claimed, you would find evidence for the infinite wisdom of the Creator. Anyone who would study His work (nature) was as legitimate a theologian as he who would study His word (the Bible). Beginning with John Ray (1691) and William Derham (1713), the study of nature became physico-theology or natural theology. It became the study of design.

Two further developments strengthened the belief in final causes. One was the increasingly strong belief that God had created the world for the sake of man. This was foreshadowed by Aristotle's statement (*Politics* 1, 8, 1256a, b), "Now if nature makes nothing incomplete and nothing in vain, the inference must be that she has made all animals for the sake of man." It was made legitimate by corresponding statements in Genesis. The other reinforcement of the belief in final causes came from manifold observations indicating ongoing changes in the world. This led to a new concept of creation. Creation was no longer seen as something that had happened instantaneously (or in six days), but as a gradual and slow process, directed by final causes, culminating in the production of man. Consistent with this modified concept of creation, G. W. Leibniz and J. G. Herder temporalized the *scala naturae,* which more and more was considered a scale of perfection. One of the foremost objectives of the writings of the physicotheologians was to demonstrate how perfectly everything in the world was designed. Since God could not have created anything that was not perfect, the world was considered the "best of all possible worlds." This was a dominant theme of that vast literature from Ray and Derham to William Paley and the Bridgewater Treatises. It dominated even Darwin's early thinking and certainly that of most of his contemporaries. Until evolution was accepted, there was no conceivable alternative to chance but "necessity," that is, God's design.

Much of the literature of natural theology is quite admirable. R. Boyle (1688), for instance, understood perfectly well that the explanation of the mechanical workings of a structure is an entirely independent endeavor from the explanation of the reason why the organ exists and what its role in the life of the organism is. Thus, he made quite clearly a distinction between proximate or immediate causations and ultimate causations. For instance, the proximate cause of sexual dimorphism in the plumage of birds is hormonal difference; the ultimate causation is sexual selection. Proximate causations could be explained mechanistically, by physical laws, but one could not explain ultimate causations without postulating a final goal or purpose (Lennox 1983).

Though the beginnings of natural theology go back to the Greeks and even the Egyptians, its period of true dominance, at least in England, lasted from the last quarter of the seventeenth century to 1859. It made little difference whether an author believed that everything in the world was governed by laws or was specifically regulated by God, because in either case God was either directly or indirectly responsible. He was the final cause of everything.

A belief in cosmic teleology fit well into the thinking of the seventeenth and eighteenth centuries. This was a period of increasing optimism, of emancipation from social and legal burdens, of conviction that better times were coming, possibly a millennium. Progress was preached not only by utopians and reformers but became the theme of philosophies, particularly of the historical-idealistic schools from Herder and Schelling to Hegel and Marx (Toulmin 1982). Nowhere else did teleology have as great an influence as in Germany. Almost all German philosophers, from Leibniz, Herder, and Kant to modern times, were teleologists to a greater or lesser extent. Kant, whose thought dominated German philosophy throughout the nineteenth century, was a teleologist (Löw 1980). As far as inanimate nature was concerned, he was a strict mechanist, but he considered all phenomena of living nature to be the product of teleological forces. Under Kant's influence, German biology in the first half of the nineteenth century was permeated with teleological thinking (Lenoir 1982). K. E. von Baer's comprehensive critique of Darwin was largely based on teleology, and so was that of the philosopher Eduard von Hartmann (1876). And the post-Darwinian teleological theory of orthogenesis had nowhere as many followers as in Germany.

Teleological thinking was strongly reinforced by the studies of the geologists and particularly by the discovery of successions of fossil faunas culminating in strata containing mammals and eventually man (Bowler 1976). It fit well with Lamarck's theory of gradual evolutionary change, this being the first genuine theory of evolution (1809). Not all progressionism in geology led to the acceptance of evolution; in fact, the majority of paleontologists from Cuvier to Agassiz thought, rather, in terms of catastrophes and subsequent more progressive new creations. Fewer and fewer authors continued to insist on the constancy of the world; most of them saw continuous change and indeed a trend toward perfection. This can be perceived in the writings of almost all authors between 1809 and 1859, even though it was expressed in various ways by authors like Meckel, Chambers, Owen, Bronn, von Baer, and Agassiz.

The general optimism of the eighteenth century received a severe jolt through the disastrous Lisbon earthquake of November 1, 1755. It induced Voltaire to satirize the Panglossian thinking of Alexander Pope and Leibniz in his *Candide*. David Hume also ridiculed claims of a harmony of nature: "Inspect a little more narrowly these living existences, the only beings worth regarding. How hostile and destructive; how insufficient all of them for their own happiness! How contemptible or odious to the spectator! The whole presents nothing but the idea of a blind nature." Kant likewise refuted the claims of natural theology. The unhappy consequences of the French Revolution contributed to the spreading of a deep pessimism. It is reflected in the thinking of Malthus and other demographers. No longer was the growth of human populations seen as one of the benefits bestowed on man by God. Rather, it was claimed that owing to limits imposed by the environment, such growth would inevitably lead to poverty, disaster, and death.

The more the studies of the naturalists progressed, the more often phenomena were found that contradicted the excellence of design. Not every organism could have been exclusively designed for its role in nature: how would this account for the existence of a limited number of well-defined types, such as mammals, birds, snakes, beetles, and so on? Rather, it was said, at the beginning relatively few archetypes were created and the laws of nature gave rise to the subsequent diversity, everything, however, had been contained in the plan of creation. Thus, indirectly, even in this thinking, diversity and adaptation were due to design (Bowler 1977).

But this revision of the design argument could not silence criticism. One asked: What is so wonderful about a parasite that tortures its victims and leads to their eventual death? Even worse, how could design be perfect if it leads to such widespread extinction, as documented by the fossil record? If the harmony of the living world, as described by natural theologians, is reflected by the mutual adaptation of organisms to one another and to their environment, and if these adaptations must be adjusted continuously to cope with the changes of the earth and with the restructuring of the faunas owing to extinction, what final causes could there be to govern all these ad hoc changes? If the environment changes, the organism has to readjust to it. But there is no necessary direction, no thought of necessary progress, and no reaching of any final goals. After evolutionary thinking had begun to spread, Matthias Schleiden (1842:61) insisted that although one can observe simple as well as complicated organisms, "it would be a totally misleading language if we would use for them the words imperfect and perfect, or lower and higher."

Natural theology, with its emphasis on design, had been virtually abandoned on the Continent by about 1800. But it continued to be strong in England, and all of Darwin's teachers and peers, particularly Sedgwick, Henslow, and Lyell, were confirmed natural theologians. This was Darwin's conceptual framework when he began to think about adaptation and the origin of species.

From Natural Theology to Natural Selection

There are many indications that when Darwin returned from the *Beagle* voyage he shared the beliefs of the natural theologians. He had wholly abandoned them, twenty-three years later, when he published the *Origin*. There is, however, not yet complete consensus in the Darwin literature under what influences and in what stages Darwin revised his interpretations. It was a peculiar period, since the British philosophers of science—Herschel, Whewell, and Mill—emphasized a rigorous scientific methodology and yet all firmly believed in final causes. They believed in laws, but God's guiding hand was needed because unguided laws would lead to random disorder (Ruse 1975b; 1979).

To what did the young Darwin attribute adaptation? Prior to 1838 his ideas on this point were rather vague. He seems to have attributed adaptation to certain laws, particularly the influence of the environment on the generative system. He still thought in terms of the design of the world. In his *Transmutation Notebooks* of this period, the most clearly teleological statement refers to dispersal: "When I show that islands would have no plants were it not for seeds being floated about,—I must state that the mechanism by which seeds are adapted for long transportation, seems to imply knowledge of whole world—if so doubtless part of system of great harmony" (D:74). Darwin's pre-1838 interpretation of evolutionary change depended on God's planning and was thus clearly a finalistic interpretation. For the Darwin of the *Transmutation Notebooks* (before September 1838), the seeming path of progression toward perfection was simply the result of certain laws that made such a development possible. All organic change, he thought, was an adaptive response to changes, however slight, in external conditions. These environmental influences induced the generative system to produce appropriate responses. This implied that God was directly involved in adaptation because only God could have made the generative system in such a way that changes in the environment would induce it to come up with an adequate response.

Yet, as Darwin's studies proceeded he discovered one phenomenon after the other that cast doubt on the perfection of adaptations (Ospovat 1981). First, he discovered all sorts of evidence for descent (called "propagation" or progression in Darwin's earlier notes), which served as a definite constraint on the absoluteness of adaptation. Then came the consideration of rudimentary or vestigial organs, which also contradicted perfect adaptation, as did the widespread occurrence of extinction. Those natural theologians, and there were others beside Darwin, who saw such inconsistencies and seeming incompatibilities with the concept of a total harmony of nature ascribed the deviations from perfect adaptation to a conflict between various laws instituted by the Creator. Organisms, said these authors, were only as perfect as is possible within the limits set by the necessity of conforming to these laws. There are, for instance, different laws required to explain the facts of structure, distribution, and succession.

Somehow such a direct reliance on eternal God-given laws for the explanation of natural phenomena must have been unsatisfactory to Darwin and in conflict with some part of his major philosophical framework. This must be the reason why he abandoned this type of thinking so speedily after he read Malthus and formulated his theory of natural selection on September 18, 1838. Natural selection gave him a purely mechanistic explanation for adaptation and for evolutionary progression. As Darwin stated in his *Autobiography* (1958:87): "The old argument of design in nature, as given by Paley, which formerly seemed to me so conclusive, fails, now that the law of natural selection has been discovered. We can no longer argue that, for instance, the beautiful hinge of bivalve shell must have been made by an intelligent being, like the hinge of a door by man. There seems to be no more design in the variability of organic beings and in the action of natural selection, than in the course which the wind blows."

After 1838 Darwin at first remained enough of a natural theologian to believe that natural selection could give him perfect adaptation. But, he abandoned this belief by the 1850s, and the *Origin* is remarkably free of any teleological language (Ospovat 1981). To be sure, the word "progress" is used ten times in this volume but almost always as a term to describe a passing of time. Only in connection with the replacement of fossil faunas of which each one seems to be "higher" than the one it has replaced does Darwin speak of a process of improvement; but he adds, "I can see no way of testing this sort of progress" (1859: 337). However, Darwin points out that there are differences in competitive ability even among living faunas. British faunal elements introduced to New Zealand are highly successful, while he doubts the reverse would be true. "Under this point of view," says Darwin, "the productions of Great Britain may be said to be higher than those of New Zealand." Yet this is not a teleological argument. The greater competitive ability of the faunal elements of Great Britain was not due to any built-in drive or final cause but simply due to the fact that the British fauna had passed through a more severe struggle for existence.

Nevertheless, the concepts of "perfect" and "perfection" continued to be popular with Darwin. In the *Origin* he used the word "perfect" 77 times, "perfected" 19 times, and "perfection" 27 times. What is remarkable, however, in these uses is how carefully Darwin makes a distinction between the product of selection and the process of perfecting. We look to his explanations in vain for a drive or tendency toward perfection. Invariably Darwin emphasizes that selection carries the evolutionary line to ever-greater perfection. This is particularly well-stated in the section of Chapter 6 with the heading "Organs of Extreme Perfection and Complication" (p. 186) which, among others, contains Darwin's well-known discussion of the evolution of eyes through natural selection. Since natural selection is not a finalistic process, Darwin now sees quite clearly that "natural selection will not necessarily produce absolute perfection; nor as far as we can judge from our limited faculties, can absolute perfection be everywhere found" (p. 206). Complete perfection, of course, is not needed, because "natural selection tends only to make each organic being as perfect as, or slightly more perfect than, the other inhabitants of the same country with which it has to struggle for existence. And we see that this is the degree of perfection attained under nature" (p. 201). There is not even a trace of a suggestion of any final cause, because perfection is simply the product of the *a posterior* process of natural selection. With the world and its biota constantly changing, perfect creation in the beginning would have been futile. "Almost every part of every organic being is so beautifully related to its complex condition of life that it seems as improbable that any part should have been suddenly produced perfect, as that a complex machine should have been invented by man in a perfect state" (*Origin,* 6th ed., pp. 58–59).

Darwin's subsequent correspondence with the American botanist Asa Gray permits us to analyze his thought even a little further. Gray, even though a rather strict creationist, accepted the importance and guiding capacity of natural selection. However, "Natural selection is not the wind which propels the vessel, but the rudder which, by friction, now on this side and now on that shapes the course" (Moore 1979:316). Variation—the wind in Gray's metaphor—was guided for Asa Gray by a divine hand. This possibility was emphatically rejected by Darwin and induced him to state his ideas, rather evidently as a direct answer to Gray, in *The Variation of Animals and Plants* (1868 II:432). Gray, however, failed to understand Darwin's argument and went even so far as to praise "Darwin's great service to natural science in bringing back to it teleology" (Gray 1876:237).

In his later years, particularly in letters to his numerous correspondents, Darwin was sometimes rather careless in his language. For instance, he referred to "the extreme difficulty or rather impossibility of conceiving this immense and wonderful universe, including man with his capacity of looking far backwards and far into futurity, as the result of blind chance or necessity." How could he have said this when the theory of natural selection had given him exactly the means to escape from the alternatives chance *or* necessity? On another occasion he said, " The mind refuses to look at this universe, being what it is, without having been designed." It is not surprising therefore that Darwin was misclassified by a number of authors who did not understand the working of natural selection. Kölliker, for instance, accused Darwin of being "in the fullest sense of the word a teleologist." And even T. H. Huxley when defending Darwin was driven to distinguish between "the teleology of Paley and the teleology of evolution" (Moore 1979:264).

Darwin was not totally alone in his rejection of finalism. Ernst

Haeckel declared emphatically that "the causes of all phenomena of nature . . . are purely mechanically acting causes, never final, the goal-directed causes" (1866 2:150). The most articulate among Darwin's supporters was August Weismann, who took up the battle for natural selection again and again and refuted the theories of Darwin's opponents (see Chapter 8). The voices of Haeckel, Weismann, F. Müller, and Darwin's naturalist friends were, however, merely cries in the wilderness, for the opposition to the mechanistic process of natural selection was almost universal. But none of his opponents truly understood natural selection, and this misunderstanding was to a large extent due to a long-standing ideological commitment to finalism. The opposition to natural selection continued up to the evolutionary synthesis and with it an open or unspoken support of finalism. For example, the geneticist T. H. Morgan, who showed his lack of understanding of natural selection even in his last book on evolution in 1932, claimed in 1910 that finalism had entered biology through natural selection because "by picking out the new variation . . . purpose enters in as a factor, for selection had an end in view," completely ignoring the randomness of variation and the statistical nature of the selection process.

Finalism as an Alternative to Natural Selection

Numerous attempts were made in the years after 1859 to replace Darwin's theory of natural selection with a superior way of achieving adaptation. The best-known of these theories are usually classified under the headings neo-Lamarckism (inheritance of acquired characters), orthogenesis (an intrinsic perfecting principle), and saltation. They all incorporated some finalistic components to a lesser or greater extent. It is not easy to report on these theories for a number of reasons. Not only are the descriptions of the postulated mechanism by which the changes are achieved usually quite vague, but the same author may support first one and then the other of these theories, or a mixture of them (Kellogg 1907; Bowler 1983, 1988). Even after the Paleyan concept of an ad hoc design of every—even the slightest—adaptation had lost all credibility, there remained a concept of a universal design of organic progression, an evolutionary reinterpretation of the temporalized scale of nature (Bowler 1977). Such a concept seemed, at first, to have a sound observational foundation. Considering that variation is random, as Darwin postulated, and considering that the number of environmental constellations is quite unlimited, one would expect a totally chaotic network of evolutionary phenomena. What one actually finds is the existence of a limited number of well-defined lineages and the possibility of arranging organisms into progressive series. This was described not only by paleontologists but also by students of living organisms, be they butterflies (Theodor Eimer) or birds (Charles O. Whitman). Variation evidently was not random but followed well-defined pathways of change. Such evolutionary trends were ascribed to an intrinsic, direction-giving force, called orthogenesis. It was described as a perfecting principle or (in German) *Vervollkommnungstrieb.*

The intrinsic nature of this force seemed to be confirmed by the fact that it was possible to establish rectilinear series not only for characters that might have been advanced by natural selection, such as increasing precision of mimicry patterns or phyletic increases in body size, but also for nonutilitarian or seemingly deleterious characters. This was an argument made particularly emphatically by Eimer (Bowler 1979). Most proposals of orthogenesis were made in strict opposition to and as alternatives to natural selection.

However, there was a group of Christian Darwinians for whom natural selection was "evidence of a directing agency and of a presiding mind" (Moore 1979). They either thought that variation as such was directive, supplying just the right material to selection, or they considered the selecting process as purposive. Clearly for them natural selection was a teleological process.

Even such an ultra-mechanist as Julius Sachs (1894) adopted Carl Naegeli's perfecting principle as the agent of all major evolutionary developments, with natural selection merely being able to improve fine-grained adaptation. Kölliker (1864) was another adherent to an autogenetic theory ascribing all evolutionary progress to "intrinsic causes," and like Naegeli he stimulated Weismann to a reply.

Evolutionary Progress without Final Causes

For some Darwinians the concept of evolutionary progress seems to have raised some embarrassing questions. How can a strictly opportunistic competitive struggle lead to progress? Darwin himself occasionally seems to have had such doubts, and they are reflected in his comment, "Never use the words higher or lower," written on the margin of his copy of the *Vestiges*. Others who also questioned progress pointed to the continued existence of the archaebacteria and other prokaryotes, to the great flourishing of the protists and lower fungi up to the present, to the parasites, and to the inhabitants of caves. None of these can be called progressive, in the sense of "higher," and yet they continue to exist and flourish. So, it was said, evolution is simply a process of specialization. And yet, who can deny that overall there is an advance from the prokaryotes that dominated the living world more than three billion years ago to the eukaryotes, with their well-organized nucleus and chromosomes as well as cytoplasmic organelles; from the single-celled eukaryotes to plants and animals with a strict division of labor among their highly specialized organ systems; and, within the animals, from ectotherms that are at the mercy of climate to the warm-blooded endotherms; and, within the endotherms, from types with a small brain and low social organization to those with a very large central nervous system, highly developed parental care, and the capacity to transmit information from generation to generation?

Attempts to define progress have been many. For Lamarck, for instance, and for many nineteenth-century authors man was clearly the most perfect organism, and all forms of life were arranged in a single column on the basis of their assumed progress toward manhood. Now we know that diversification is the most characteristic attribute of evolution and that life not only very early split within the prokaryotes into eubacteria and archaebacteria, but the eukaryotes after their origin quickly gave rise to the protists, and to the kingdoms of fungi, plants, and animals. Literally thousands of distinct phyletic lines developed within each of these kingdoms, most of them not in the slightest tending toward the characteristics of man. Neither can the dominance of a group on earth be considered the criterion of progress. On that basis the vascular plants would have to be considered more dominant than man and even the insects. And man's ancestors until less than ten thousand years ago were anything but dominant on earth.

Structural complexity is sometimes mentioned as a sign of progress, but trilobites and placoderms would seem to have been more complex in structure and perhaps more specialized than modern man. Huxley (1942) considered emancipation from the environment an important index of progressiveness, and in that criterion man certainly ranks higher than any other organism. However, is independence from the environment truly an index of progressiveness?

When discussing evolutionary progress one seems to be quite unable, since one is a member of the human species, to get away from criteria that would give man supremacy. However, there are two criteria of progressiveness that would seem to have a considerable amount of objective validity. One of these is parental care (promoted by internal fertilization), which provides the potential for transferring information nongenetically from one generation to the next. And the possession of such information is of course of considerable value in the struggle for existence. This information transfer generates at the same time a selection pressure in favor of an improved storage system for such remembered information, that is, an enlarged central nervous system. And, of course, the combination of postnatal care and an enlarged central nervous system is the basis of culture, which together with speech, sets humans quite aside from all other living organisms. However, even if we would designate the acquisition of these capacities as evidence for evolutionary progress, it would not strengthen the case for final causes, since these developments were clearly achieved through natural selection.

Whether one is looking at the highest mammals and birds, the social insects, the orchids, or giant trees, it has seemed inconceivable to some students of evolution that the slow struggle for existence among individuals of a species could account for the enormous evolutionary progress observed in so many phyletic lines. To see all evolution simply as the result of competition among the individuals in a population is indeed simplistic, because superimposed on this individual selection is a process traditionally referred to as species selection, although perhaps a better term would be species replacement or species succession. An individual organism competes not only with members of its own species but struggles for existence also against members of other species. And this process is probably the greatest source of evolutionary progress. Each newly formed species, if it is evolutionarily successful, must represent, in some way, evolutionary progress. Darwin explained this as follows: "But in one particular sense the more recent forms must, on my theory, be higher than the more ancient; for each new species is formed [that is, has become successful] by having had some advantage in the struggle for life over other and preceding forms" (1859: 337). When the competition among individuals of different species leads to or at least contributes to the extinction of one of the competing species, it is a case of species replacement. That competition among species could lead to the extinction of one of the competitors was of course already known to Lyell and other pre-evolutionary authors.

A new species will be successful in the struggle for existence over a previously existing species only if it has made some, even the smallest, evolutionary invention. This might be an improvement in its digestive physiology or its nervous system or its lifestyle or any other of the countless ways by which the so-called "higher" organisms differ from the lower ones. Thus, the Darwinian mechanisms of variation and selection, of speciation and extinction, are fully capable of explaining all macroevolutionary developments, whether specializations, improvements, or other innovations. And none of this requires any finalistic agent.

Close study of evolutionary progress shows that its characteristics are not compatible with what one would expect from a process guided by final causes. Progressive changes in the history of life are neither predictable nor goal-directed. The observed advances are haphazard and highly diverse. It is always uncertain whether newly acquired adaptations are of permanent value. Who at the beginning of the Cretaceous would have predicted the total extinction by the end of the period of that flourishing taxon the dinosaurs? Episodes of stasis alternate with episodes of precipitous evolutionary change. Evolutionary trends are rarely rectilinear for any length of time, and when such rectilinearity occurs it can usually be shown to be due to built-in constraints.

All the evolutionary phenomena and aspects of evolutionary progress that were considered as irrefutable proof of teleology by earlier generations can now be shown to be entirely consistent with natural selection. Phenomena that are due to a chain of historical events cannot be ascribed to simple laws and can therefore not be proven in the same way as can phenomena studied in the physical sciences. However, they can be shown to be consistent with the findings of genetics and with the theory of natural selection in its modern sophisticated form. No one has refuted the finalistic thesis of evolution more convincingly than Simpson (1949; 1974). He pointed out that each evolutionary line goes its own way, and evolutionary progress can be defined only in terms of that particular lineage. Nothing seemed more progressive in the geological past than the ammonites and the dinosaurs, and yet both taxa became extinct. On the other hand many evolutionary lines have displayed no evidence of progress in hundreds or thousands of millions of years, and yet they have survived to the present day, as the archaebacteria and other prokaryotes. Progress thus is not at all a universal aspect of evolution, as it ought to be if evolution were generated by final causes.

The Decline, if not Demise, of Finalism

By the time of the evolutionary synthesis of the 1940s (see Chapter 9), virtually no evolutionary biologist, in fact no competent biologist, was left who still believed in any final causation of evolution. The few biologists who still did so either had theological commitments, like Teilhard de Chardin, or were unaware of the developments of biology in the twentieth century, like Comte de Nouy.

Final causes, however, are far more plausible and pleasing to a layperson than the haphazard and opportunistic process of natural selection. For this reason, a belief in final causes has had a far greater hold outside of than within biology. Almost all philosophers, for instance, who wrote on evolutionary change in the one hundred years after 1859 were confirmed finalists, from Whewell, Herschel, and Mill to Henri Bergson in France, who postulated a metaphysical force, *élan vital,* which, even though he disclaimed its finalistic nature, could not have been anything except a final cause, considering its effects. Whitehead, Polanyi, and many lesser philosophers were also finalistic. Throughout this period there have been exceptions, the most noteworthy perhaps being the German philosopher Sigwart, who as early as 1881 provided a remarkably modern treatment of the problems of teleology.

Modern philosophers—that is, those who have published since the evolutionary synthesis—have, on the whole, refrained from invoking final causes when discussing evolutionary progress. Apparently they fully accept the explanation provided by the evolutionary synthesis. When they do discuss teleology, like Morton Beckner or Ernest Nagel, they deal with adaptation and with "teleological systems." Finalism is no longer part of any respectable philosophy. One last vigorous attack on finalism was Monod's book *Chance and Necessity* (1970). But Monod failed to understand the explanatory power of natural selection and opted for pure chance as having been responsible for the phenomena of nature. Such Epicureanism, however, is only rarely encountered in modern times.

The reason why the controversy about the validity of teleological thinking has been so indecisive has finally become evident in recent years: the designation "teleological" has been applied to four quite different natural phenomena. Three of these can be explained by science, while the fourth one, an explanatory postulate for certain phenomena, has not been substantiated.

(1) Many seemingly end-directed processes or movements in inorganic nature are the simple consequences of natural laws. The falling of a stone (due to gravity) or the cooling of heated pieces of metal (owing to the first law of thermodynamics) are examples of such *teleomatic* processes, as such law-directed processes are called.

(2) Processes in living organisms—as well as their behavior—that owe their goal-directedness to the operation of an inborn or acquired program are called *teleonomic*. This includes all changes in ontogenetic development as well as end-directed behavioral activities. Such processes can be analyzed strictly scientifically, since the end-point or goal is already contained in the program.

(3) Adapted systems, like the heart, which pumps blood, or the kidneys, which eliminate by-products of protein metabolism, and which seem to work toward a goal have also been called teleological. An organism has hundreds, if not thousands, of such adapted systems, from the molecular level up to the organism as a whole, all of them acquired during the evolution of its ancestors and continuously fine-tuned by natural selection. These systems have the capacity for teleonomic behavior, but, being stationary, are not themselves goal-seeking.

(4) From the Greeks on, there was a widespread belief that everything in nature and its processes has a purpose, a predetermined goal. And these processes would lead the world to ever-greater perfection. Such a teleological worldview was held by many of the great philosophers. Modern science, however, has been unable to substantiate the existence of such a *cosmic teleology*. Nor have any mechanisms or laws been found that would permit the functioning of such a teleology. The conclusion of science has been that final causes of this type do not exist.

CHAPTER SIX

Darwin's Path to the Theory
of Natural Selection

W HEN WE SPEAK of Darwinism today, we mean evolution by natural selection. The meaning of natural selection, its limits, and the processes by which it achieves its effects are now the most active areas of evolutionary research.

The fifth one of Darwin's great evolutionary theories was his most daring, most novel. It dealt with the *mechanism* of evolutionary change and, more particularly, how this mechanism could account for the seeming harmony and adaptation of the organic world. It attempted to provide a natural explanation in place of the supernatural one of natural theology. In that respect Darwin's theory was unique; there was nothing like it in the whole philosophical literature from the pre-Socratics to Descartes, Leibniz, or Kant. It replaced teleology in nature with an essentially mechanical explanation.

To judge from his writings, Darwin had a much simpler concept of natural selection than the modern evolutionist does. For him there was a steady production of individuals, generation after generation, some of whom were "superior" in having a reproductive advantage. For Darwin selection was essentially a single-step process, the conveying of reproductive success. The modern evolutionist agrees with Darwin that the individual is the target of selection; but we now also realize that natural selection is actually a two-step process, the first one consisting of the production of genetically different individuals (variation), while the survival and reproductive success of these individuals is determined in the second step, the actual selection process. Although I have called the theory of natural selection Darwin's fifth theory, it is actually itself a small package of theories. This includes the theory of the perpetual existence of a reproductive surplus, the theory of a continuing availability of great genetic variability, the theory of the heritability of individual differences, the theory that mere reproductive superiority is selected for (sexual selection), and several others.

The question concerning the conceptual sources of Darwin's theory of natural selection is still highly controversial. A favorite interpretation among historians has always been that it was a manifestation of the thinking of upper-class England in the first half of the nineteenth century (consistent with empiricism, mercantilism, industrial revolution, poor laws, and so forth). Darwin's admission that reading Malthus had given him the crucial insight seemed to provide a powerful confirmation of this "external causation." The evolutionists, by contrast, have favored an interpretation based on "internal causation," relying on Darwin's insistence that his familiarity with the practices of the animal breeders had provided him with the decisive evidence. The rediscovery of Darwin's notebooks covering the one and a half years prior to the date of his "conversion" has provided us with a great amount of new information, but—although narrowing down our options—it still permits conflicting interpretations. What I present here is anything but the last word in a still-ongoing controversy. It will require further research before the remaining disagreements can be removed (Hodge and Kohn 1985).

Darwin had returned to England from the voyage of the *Beagle* in October 1836. While working on his bird collections, and particularly through discussions with the ornithologist John Gould, Darwin became an evolutionist, apparently in March 1837 (Sulloway 1982b). Certainly by July 1837 he had firmly accepted evolution by common descent. His new interpretation of the world consisted not only in replacing a static or steady-state world by an evolving one but also, and more important, in depriving man of his unique position in the universe and placing him into the stream of animal evolution. Darwin, after this date, never questioned the fact of evolution, even though he continued for another twenty years to collect supporting evidence. Yet, the causes of evolution were at first a complete mystery to him.

For a year and a half Darwin speculated incessantly, developing and then again rejecting one theory after the other (Kohn 1975), until he finally had a decisive illumination on September 28, 1838. In his autobiography he describes it as follows (Darwin 1958: 120):

> Fifteen months after I had begun my systematic enquiry, I happened to read for amusement Malthus on Population, and being well prepared to appreciate the struggle for existence which everywhere goes on, from long-continued observation of the habits of animals and plants, it at once struck me that under these circumstances favorable variations would tend to be preserved, and unfavorable ones to be destroyed. The result of this would be the formation of new species. Here, then, I had at last got a theory by which to work.

It was the theory later called by Darwin the theory of natural selection. It was a most daring innovation, since it proposed to explain by natural causes, mechanically, all the wonderful adaptations of living nature hitherto attributed to "design."

Darwin makes it sound as though the concept of natural selection was simplicity itself. But his memory deceived him. His autobiography was written almost forty years later (in 1876), largely for the benefit of his grandchildren, and was replete with characteristically Victorian self-denigrations. Darwin had forgotten what a complex shift in four or five major concepts had been required to arrive at the new theory. He probably never fully realized himself how unprecedented his new concept was and how totally opposed to many traditional assumptions.

Indeed, the concept of natural selection was so strange to Darwin's contemporaries when he proposed it in the *Origin of Species* that only a handful adopted it. Nearly three generations passed before it became universally accepted even among biologists. Among nonbiologists the idea is still unpopular, and even those who pay lip service to it often reveal by their comments that they do not fully understand the working of natural selection. Only when one is aware of the complete unorthodoxy of this idea can one appreciate Darwin's revolutionary intellectual achievement. And this poses a powerful riddle: How could Darwin have arrived at an idea which not only was totally at variance with the thinking of his own time but was so complex that even now, one and a half centuries later, it is widely misunderstood in spite of our vastly greater understanding of the processes of variation and inheritance?

Darwin's own version (in his autobiography) was that contemplation of the success of animal breeders in producing new breeds had provided him with the clue for the mechanism of evolution and was thus the basis for his theory of natural selection. We know that this is a vast oversimplification—a revision of our thinking which we owe to the rediscovery of Darwin's notebooks. In July 1837 he had started to write down all the facts as well as his own thoughts and speculations "which bore in any way on the variation of animals and plants under domestication and [in] nature." Even though he later cut out occasional pages, to use them for his book manuscripts, Darwin never discarded these notebooks, and they were rediscovered in the 1950s among the Darwin papers at the Cambridge University Library (Barrett et al. 1987). Darwin's day-by-day records throw an entirely new light on the development and the changes in his thought during the period from July 1873 to September 28, 1838, when the theory of natural selection was conceived.

One fact, the importance of which has not been reduced by the recent discoveries, is the impact of Darwin's reading of Malthus. The interpretation of the Malthus episode, however, has become the subject of considerable controversy among the Darwin scholars. According to some of them—de Beer (1961) and S. Smith (1960), and, to a lesser extent, Gruber (1974) and myself—it was merely the culmination in the gradual development of Darwin's thinking, a little nudge that pushed Darwin across a threshold he had already reached. According to others—Limoges (1970) and

FIGURE I

Darwin's Explanatory Model of Evolution through Natural Selection

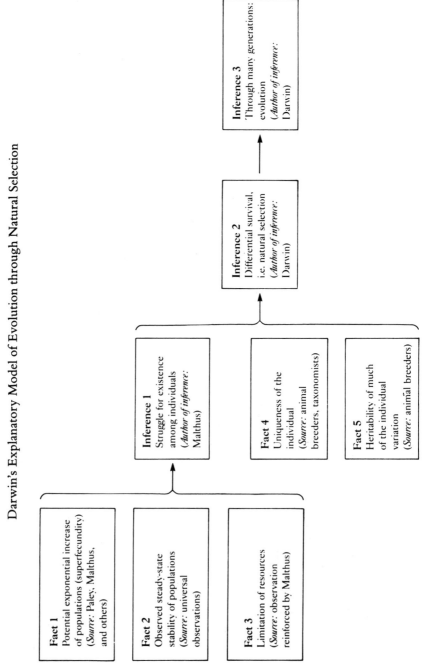

Kohn (1980), for example—it constituted a rather drastic break, almost equivalent to a religious conversion. Which of these two interpretations is nearer to the truth?

There are essentially two methods by which we can try to find an answer. Either we can attempt to analyze all the entries in the notebooks, in a chronological sequence, or we can try to reconstruct Darwin's explanatory model of natural selection and then study separately the history of each of its individual components. My own choice is in favor of the second method, placed in a chronological framework, although both methods are necessary for a full understanding. The first method attempts to reconstruct the trials and errors of Darwin's gradual approach, as reflected in successive entries in the notebooks. It also examines tentative ideas that he later rejected. Gruber, Kohn, and, in part, Limoges have favored this method.

Darwin's Explanatory Model

What were the components of Darwin's explanatory model? For my analysis I have found it most convenient to recognize five facts and three inferences (see Figure 1). I shall attempt to determine, first, at what time Darwin became aware of these five facts, then at what time he made the three inferences, and whether or not these inferences had already been made previously and could be found in the literature.

The five facts were already widely known before the Malthus episode, not only to Darwin but to his contemporaries, only a single one of whom, A. R. Wallace, used them in exactly the same way as Darwin. Merely having these facts obviously was not enough. They had to be related to one another in a meaningful manner; that is, they had to be placed in an appropriate conceptual background. In other words, Darwin had to be intellectually prepared to see the connections among these facts.

This leads us to the most interesting but also the most difficult question: What had been going on in Darwin's mind in the one and a half years prior to the Malthus episode? All the indications are that it was a period of unprecedented intellectual activity in Darwin's life. Precisely what the changes in Darwin's thinking were and how they were connected with one another has not yet been investigated nearly as fully as it deserves. Gruber and Kohn have examined this problem more carefully than anyone else, but the Darwin correspondence of that period and other manuscript materials that have not yet been thoroughly analyzed are bound to provide new insights. My own tentative conclusions may, therefore, turn out to be incorrect. However, my reading suggests to me that Darwin's beliefs changed moderately or drastically in four areas, which I shall simply list here and then discuss in the context of Darwin's model.

(1) *The gradual replacement of the assumption that all individuals of a species are essentially alike by the concept of the uniqueness of every individual.* As discussed previously, the belief that the observed variability of phenomena reflects a limited number of constant, discontinuous essences was gradually replaced by population thinking, a belief in the reality of variation within a population and in the importance of these individual differences. Most of Darwin's earlier statements on species and varieties were strictly typological. It is my impression that they became more popula- tional as Darwin delved deeper into the literature of the animal breeders and also later as a result of his work on the barnacles (Ghiselin 1969).

(2) *A shift from soft toward hard inheritance.* In his earlier state- ments Darwin seemed to assume that most, if not all, inheritance was "soft." He assumed that the material basis of inheritance is not unchangeably constant but can be modified by use and dis- use, by physiological activities of the body, by a direct influence of the environment on the genetic material, or by an inherent tendency to progress toward perfection, and that these environ- mentally induced changes could be passed along to the next gen- eration. This theory is called the inheritance of acquired charac- ters. The growth of his population thinking, with its increasing stress on genetic differences among individuals, indicates a grow- ing awareness of the need to postulate "hard" inheritance—that

is, inheritance that is not affected directly by environmental factors.

(3) A changing attitude toward the balance of nature. Darwin began to believe that the balance of nature is dynamic rather than static, and he began to ask whether the balance is maintained by benign adjustments or by constant war.

(4) A gradual loss of his Christian faith. Darwin lost his faith in the years 1836–1839, much of it clearly prior to the reading of Malthus. In order not to hurt the feelings of his friends and of his wife, Darwin often used deistic language in his publications, but much in his notebooks indicates that by this time he had become a "materialist" (more or less equivalent to an atheist; see Chapter 2).

These four changes in Darwin's thinking are to some extent interconnected. Since they were largely unconscious, they are usually reflected in Darwin's notebooks only in subtle changes of wording, and there is considerable leeway in possible interpretation. However, keeping these four points in mind will sharpen our awareness of possible changes in Darwin's thinking in the years prior to reading Malthus, while we make a point-by-point analysis of Darwin's explanatory model.

The Struggle for Existence

When recording his reaction to reading Malthus, Darwin makes it quite clear that it was not Malthus's general attitude that had acted as a catalyst on his thoughts but one particular sentence, for he says that "yet until the one sentence of Malthus no-one clearly perceived the great check amongst men" (D:135). De Beer (1963:99) succeeded in determining what Malthus's crucial sentence was: "It may safely be pronounced, therefore, that the population, when unchecked, goes on doubling itself every twenty-five years, or increases in a geometrical ratio." From then on, Darwin stressed that it was Malthus's demonstration of the exponential increase of populations that was decisive in his discovery of the importance of natural selection (Fact I).

Yet there is a puzzling difficulty. Why did it take Darwin so long to recognize the evolutionary significance of the Malthusian principle? The prodigious fertility of animals and plants had been pointed out by many of Darwin's favorite and most frequently read authors, like Erasmus Darwin, Charles Lyell, Alexander von Humboldt, and William Paley. Furthermore, Malthus's principle was widely discussed in the essay literature of the period. Why then did this suddenly impress Darwin so profoundly on September 28, 1838?

Four reasons might be suggested—the first one being, as pointed out by Gruber (1974), that Darwin had learned on the three preceding days (between September 25 and 27) of the unbelievable fertility of protozoans by reading Ehrenberg's work on the subject. This quite likely primed Darwin's receptivity for Malthus's thesis. The second reason is that when Malthus applied the principle to man, a species with relatively few offspring, Darwin suddenly realized that a potentially exponential increase of a population was entirely independent of the actual number of offspring of a given pair. The third reason is that the Malthus episode came at a time when population thinking had begun to mature in Darwin's mind. The fourth reason, suggested by Ruse (1979: 175), is that the numerical formulation suggested by Malthus seemed to satisfy the mathematical requirements of such Newtonians as Herschel.

The second fact in Figure 1—population stability—was not in the slightest bit controversial. No one questioned that the number of species and, aside from temporary fluctuations, the number of individuals in every species maintained a steady-state stability. This is implicit in the concept of plenitude of the Leibnizians and in the harmony-of-nature concept of the natural theologians. If there is an extinction, it is balanced by speciation, and if there is high fertility it must be counterbalanced by mortality. In the end everything adds up to a steady-state stability.

The third fact—limitation of resources—again was not at all controversial, being very much part of the balance-of-nature concept of natural theology, so dominant in England in the first half of the nineteenth century.

Darwin's first great inference, derived from these three facts, was that exponential population growth combined with a fixed amount of resources would result in a fierce struggle for existence. We must ask if this inference was original with Darwin, and if so, what part of it did he owe to Malthus? This is perhaps the most controversial question raised by the analysis of the selection theory. The main difficulty is that the term "struggle for existence" and similar synonymous terms were used in different senses by different authors.

Before we can analyze them, we must deal with one other concept, the idea of a perfect balance of nature, an idea prevalent in the eighteenth century: nothing in nature is too much, nothing too little, everything is designed to fit with everything else. Rabbits and hares have lots of young because food must be available for foxes and other carnivores. The whole economy of nature forms a harmonious whole that can in no way be disturbed. This is why Lamarck, who was very much an adherent of this concept, could not conceive of extinction. Cuvier likewise had adopted the idea of perfect balance, as shown in correspondence with his friend Pfaff. He transferred the same concept to the structure of an organism, which he visualized as a "harmonious type" in which nothing could be changed. Everything in such a complex system is so perfect that any change would lead to deterioration.

This type of thinking was still dominant in Darwin's day, not only among the natural theologians of England but also on the Continent. Indeed, one can find a number of entries in Darwin's notebooks that seem to reflect this kind of thinking. But was Darwin still a wholehearted supporter of the concept of a harmonious balance of a benign nature? This is a very important question because it affects the interpretation of what Darwin understood under the term "struggle for existence."

For us moderns the term means a fierce fight with no holds barred. But for the natural theologians the struggle for existence was a beneficial feedback device, the function of which was to maintain the balance of nature. It is, as Herder (1784) called it, "the balance of forces which brings peace to the creation." Linnaeus (1781) devoted an entire essay to the "Police of Nature" and emphasized that "those laws of nature by which the number of species in the natural kingdoms is preserved undestroyed, and their relative proportions kept in proper bounds are objects extremely worthy of our attentive pursuit and researches." Lamarck expressed similar sentiments.

Was this benign interpretation of the struggle for existence unanimous? Unfortunately, even today we have no reliable analysis that would give us an answer to this question. My impression is, however, that as the interaction of predators and prey, of parasites and their victims, the frequency of extinction, and the struggles of competing species became better known, the struggle for existence was more and more recognized as a "war" or fight, a struggle for survival, "red in tooth and claw" as Tennyson later expressed it. Bonnet (1781) and de Candolle (1820) emphasized that this war among species consisted not merely of predator–prey relationships but of competition for any and all resources. However, it was not at all appreciated how fierce this struggle is, and Darwin admits that "even the energetic language of de Candolle does not convey the warring of the species as [convincingly as does the] inference from Malthus."

Nevertheless, it is highly probable that Darwin had been gradually conditioned by his reading to a far less benign interpretation of the struggle for existence than that held by the natural theologians. The mere fact that Darwin had adopted evolution must have made him aware of the frequency of extinction and of the unbalances and adaptational lags caused by evolutionary changes. From Aristotle to the natural theologians it was considered axiomatic that a belief in a harmonious universe and perfect adaptation in nature, or in a Creator continuously active in correcting imperfections and imbalances, was incompatible with a belief in evolution. By necessity, accepting evolutionary thinking undermined a continued adherence to a belief in a harmonious universe.

Struggle among Species or Individuals?

Of far greater importance is a second question: Between whom does the struggle for existence take place? This question allows two drastically different answers. In the entire essentialistic literature the struggle is considered to take place among species. The balance of nature is maintained by this struggle, even if it occasionally causes the extinction of a species. This is the interpretation of the struggle for existence in the literature of natural theology, up to de Candolle and Lyell, and is the major emphasis of Darwin's notebooks up to the Malthus reading. The main function of this struggle is to correct disturbances in the balance of nature, but it can never lead to changes; on the contrary, it is a device to preserve a steady-state condition. As such it continued even after 1838 to be an important component of Darwin's thinking, particularly in his biogeographic discussions (such as determination of species borders).

Only when one applies population thinking to the struggle for existence can one make the crucial conceptual shift to recognizing a struggle for existence among individuals of a single population. This, as Herbert (1971) was the first to recognize clearly, was Darwin's decisive new insight resulting from his reading of Malthus, although Mayr (1959b) and Ghiselin (1969) had previously pointed out the populational nature of selection. If most individuals of every species are unsuccessful in every generation, then there must be a colossal competitive struggle for existence among them. It was this conclusion that made Darwin think at once of various other facts that had been slumbering in his subconscious but for which, up to that moment, he had had no use.

Darwin's reading of Malthus was dramatic and climactic, and it does not matter whether one interprets it as a complete reversal of Darwin's thinking or whether one believes that "the evidence suggests that the change in choice of unit was a protracted process, stretched over a year or more, and linked to other aspects of his thought" (Gruber 1974). I myself hold with the latter view, because the capacity to be able to interpret the Malthus statement on exponential growth of populations and to apply it to individuals requires population thinking, and this Darwin had been gradually acquiring during the preceding year and a half. That everything came to a dramatic climax on September 28, 1838, however, is beyond question.

The whole concept of competition among individuals would be irrelevant if all these individuals were typologically identical—if they all had the same essence. Competition does not become meaningful in an evolutionary sense until a concept has developed that allows for variability among the individuals of the same population. Each individual may differ in the ability to tolerate climate, to find food and a place in which to live, to find a mate, and to raise young successfully. The recognition of the uniqueness of every individual and the role of individuality in evolution is not only of the utmost importance for an understanding of the history of biology, but it is one of the most drastic conceptual revolutions in Western thought (Fact 4).

There is little doubt that this concept—which we call "population thinking"—received an enormous boost in Darwin's mind through his reading Malthus at that right moment. Yet, curiously, when we go through Malthus's writings we find no trace of population thinking. There is nothing whatsoever even faintly relating to the subject in those early chapters of Malthus that gave Darwin the idea of exponential growth. There is a reference to animal breeding in chapter 9, but there the subject is introduced to prove exactly the opposite point. After referring to the claims of the animal breeders, Malthus states, "I am told that it is a maxim among the improvers of cattle that you may breed to any degree of nicety you please, and they found this maxim upon another, which is, that some of the offspring will possess the desirable qualities of the parents in a greater degree." He then pro-

J. S. Henslow (1796–1861), Darwin's botany professor at Cambridge University, who procured for Darwin the invitation to join the *Beagle* voyage

The *H.M.S. Beagle,* in the Straits of Magellan, 1833

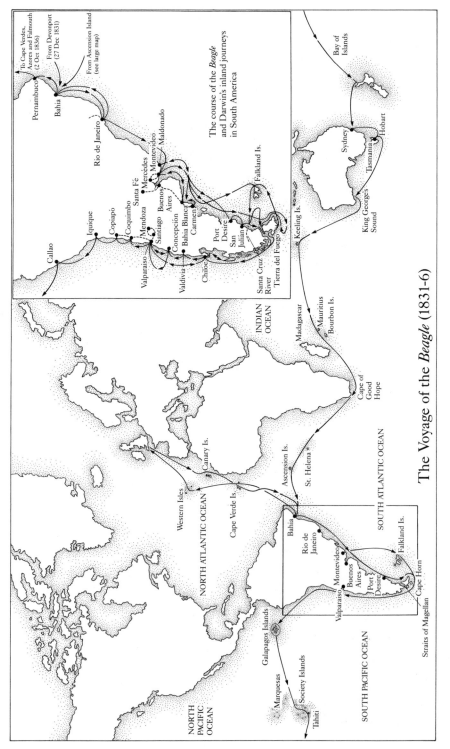

The route of the voyage of the *Beagle*, 1831–1836

The naturalist Jean Baptiste Lamarck (1744–1829), the first person to adopt a consistent theory of gradual evolutionary change

Charles Lyell (1797–1875), geologist, to whose theory of uniformitarianism Darwin owed much of his thinking about evolutionary change

The ornithologist John Gould (1804–1881), who informed Darwin in March 1837 that the specimens of mockingbirds he had collected on three islands in the Galapagos were three distinct species

Emma Wedgwood Darwin (1808–1896) in 1840, Charles Darwin's wife and first cousin, and daughter of the famous potter Josiah Wedgwood II

Charles Darwin in 1840

The botanist Joseph Dalton Hooker (1817–1911), Darwin's friend and supporter, who in 1858, with Lyell, presented Darwin's and Wallace's findings to the Linnean Society of London

The morphologist and paleontologist Richard Owen (1804–1892), once Darwin's friend, who viciously attacked the *Origin* and whose statements Huxley attacked with equal vigor

Alfred Russel Wallace (1823–1913), co-discoverer with Darwin of the theory of evolution through natural selection and opponent, with Weismann, of the theory of the inheritance of acquired characters

The morphologist, physiologist, and embryologist Thomas Henry Huxley (1825–1895), self-anointed as "Darwin's bulldog" for his energetic defense of Darwin and the theory of descent

Above: The botanist Asa Gray (1810–1888), Darwin's most important supporter in America and a devout Christian who succeeded in reconciling natural selection with a belief in a personal god

Below: The Swiss-born naturalist Louis Agassiz (1809–1873), an outstanding ichthyologist and specialist of other groups of organisms, who played a prominent role in the spread of natural history studies in America; a devout creationist, Agassiz characterized the *Origin* as "a scientific mistake, untrue in its facts, unscientific in its method, and mischievous in its tendencies"

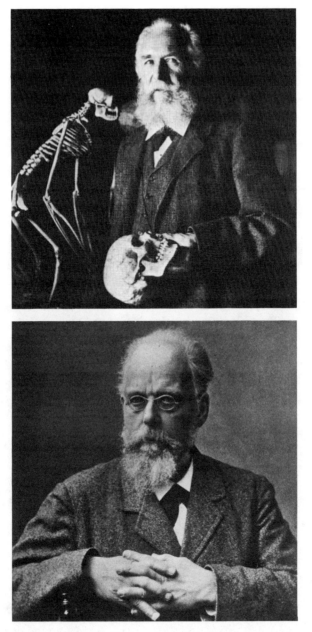

Above: The biologist Ernst Haeckel (1834–1919), an enthusiastic supporter and popularizer of Darwinism in Germany, who specially emphasized the study of phylogeny

Below: August Weismann (1834–1914), the nineteenth century's greatest evolutionist after Darwin and a staunch supporter of natural selection; he was responsible for the demise of the theory of acquired characters and laid the foundations of genetics

Charles Darwin on the veranda at Down House

duces all sorts of facts and reasons why this cannot possibly be the case, leading him to the conclusion that "it cannot be true, therefore, that among animals, some of the offspring will possess the desirable qualities of the parents in a greater degree; or that animals are indefinitely affectable" (Malthus 1798:163).

Where, then, did Darwin get his population thinking, since he evidently did not get it from Malthus? In his autobiography and in various letters Darwin emphasized again and again that he had been mentally prepared for the Malthus principle by studying the literature of animal breeding. Recent commentators such as Limoges and Herbert have insisted that this must be a lapse of Darwin's memory because there is very little about animal breeding in Darwin's notebooks until about three months after the Malthus reading. For myself, I am convinced that Darwin's own presentation is nevertheless essentially correct.

If we ask ourselves what Darwin would be likely to enter in his notebooks, we would certainly say new facts or new ideas. Hence, since it was not a new subject, animal breeding surely would not qualify. Darwin's best friends at Cambridge University were the sons of country squires and of owners of estates. They were the "horsy set," riding or hunting with dogs on every occasion (Himmelfarb 1959). All of them to a greater or lesser degree were interested in animal breeding. They must have argued a great deal among themselves about Bakewell and Sebright and the best methods of breeding and improving dogs, horses, and livestock.

How else—other than that it had a great interest for him—can one explain that Darwin, in the excessively busy period after the return of the *Beagle,* devoted so much of his time to studying the literature of the animal breeders? To be sure, Darwin's primary interest was in the origin of variation, but in the course of his reading Darwin could not help absorbing the important lesson from the breeders—that every individual in the herd was different from every other one and that extreme care had to be used in selecting the sires and dams from which to breed the next generation. I am quite convinced that it is no coincidence that Darwin studied the literature of the animal breeders so assiduously exactly during the six months before reading Malthus (Ruse 1975a).

It was not the process of selection but the fact of the differences among individuals that Darwin remembered when suddenly becoming aware of the competition among individuals, of the struggle for existence among individuals. Here we have the fortuitous coming together of two important concepts—excessive fertility and individuality—which jointly provide the basis for an entirely new conceptualization.

Variation can be of evolutionary significance—that is, it can be selected—only if at least part of it is heritable (Fact 5). Like the animal breeders from whom he got so much of his information, Darwin took this heritability of characters completely for granted. And this assumption can be held quite independently of one's assumptions concerning the nature of the genetic material and of the origin of new genetic factors. Darwin's ideas on these subjects were quite confused, but he did know a number of things by observation. He knew that in asexual reproduction the offspring are identical with the parent, while in sexual reproduction offspring are different from the parent and from one another. Furthermore, he knew that each offspring had a mixture of the characters of both parents. On the whole Darwin treated genetic variation as a "black box." As a naturalist and reader of the animal breeding literature, he knew that variation was always present, and this is all he *had* to know. He was also convinced that the supply of variation was renewed in every generation and thus was always abundantly available as raw material for natural selection. In other words, a correct theory of genetics was *not* a prerequisite for the theory of natural selection.

The Path to Discovery

The next question we have to answer is how Darwin arrived at the actual concept of natural selection on the basis of the stated five facts and his first inference. In his autobiography (1958: 118–

120) Darwin stressed that he "collected facts on a wholesale scale, more especially with respect to domesticated productions, by printed enquiries, by conversation with skillful breeders and gardeners, and by reading . . . I soon perceived that selection was the key-stone of man's success in making useful races of animals and plants. But how selection could be applied to organisms living in a state of nature remained for some time a mystery to me." In 1859 he wrote to Wallace, "I came to the conclusion that selection was the principle of change from the study of domestic productions; and then, reading Malthus, I saw at once how to apply this principle." To Lyell he wrote, with reference to Wallace's theory, "We differ only [in] that I was led to my views from what artificial selection had done for domestic animals." Traditionally, these statements were accepted by Darwin students as a correct representation of the facts.

This interpretation of Darwin's path to the concept of natural selection has been challenged in recent years in the wake of the discovery of Darwin's notebooks, for the same reason that researchers have come to doubt the sources of Darwin's population thinking. Limoges and Herbert point out that in the first three notebooks Darwin nowhere refers to selection or to the selective activities of animal breeders, particularly in the production of new domestic races. They claim that Darwin was interested in domestic animals only because he hoped to find evidence concerning the occurrence of variations and the mechanisms of their production, matters that are difficult to study in wild populations.

It is true that the term "selection" does not occur in Darwin's notebooks; it is first found in his 1842 sketch in the words "natural means of selection" (F. Darwin 1909:17). Darwin here refers to artificial selection by the term "human selection." Actually, in the notebooks Darwin not infrequently refers to the process of selecting, but he uses a different term—"picking."

I am willing to grant to the recent critics that there is no evidence in the notebooks of a simple application to the evolutionary process of the analogy between selection by man and selection by nature. This is quite evident when one reads the crucial entry in the notebooks on September 28, 1838 (here reproduced in the original telegraph style):

> Take Europe on an average every species must have same number killed year with year by hawks, by cold etc.—even one species of hawk decreasing in number must affect instantaneously all the rest.—The final cause of all this wedging, must be to sort out proper structure, and adapt it to changes—to do that for form, which Malthus shows is the final effect (by means however of volition) of this populousness on the energy of man. One may say there is a force like a hundred thousand wedges trying [to] force every kind of adapted structure into the gaps in the economy of nature, or rather forming gaps by thrusting out the weaker ones.

The metaphor here is "wedging," not "selecting." Thus it appears that the arguments of the critics have considerable validity. However, the analogy between artificial selection and natural selection is not necessary for Darwin's conclusions. Inference 1 and Fact 4 automatically result in Inference 2 (natural selection). It is quite likely that Darwin did not see the obvious analogy between artificial and natural selection until some time after the Malthus reading. Yet, I have little doubt that the copious reading Darwin had done in the field of animal breeding had prepared his mind to appreciate the role of the individual and its heritable qualities. Indeed, I am convinced, with Ruse, that the many years during which Darwin had been exposed to the ideas of the animal breeders had preconditioned his mind to appreciate the importance of the Malthus principle. This dormant knowledge was actualized under the impact of reading Malthus.

The natural selection of individuals with particular heritable qualities, continued over many generations, automatically leads to evolution, as in Inference 3. In fact, this process is sometimes used as the definition of evolution. In this connection it must be emphasized once more that Darwin's inference is exactly the opposite of that of Malthus, who had denied that "some of the offspring will possess the desirable qualities of the parents in a greater degree." Indeed, Malthus used his entire argument as a refutation of the thesis of Condorcet and Godwin of human perfectibility. The Malthusian principle, dealing with populations of essentialistically identical individuals, causes only quantitative, not qualitative, changes in populations (Limoges 1970). The frequently upheld thesis that it was the social-sciences message of Malthus that was responsible for Darwin's new insight has been convincingly refuted by Gordon (1989).

How Great Was Darwin's Debt to Malthus?

That the Malthus reading acted as a catalyst in Darwin's mind in producing the theory of natural selection cannot be disputed and was emphasized by Darwin himself again and again. However, when we analyze the components of the theory, as we have just done, we find that it is primarily the insight that competition is among individuals rather than species that is clearly a Malthusian contribution. To be sure, this in turn led Darwin to a reevaluation of other phenomena, such as the nature of the struggle for existence, but only as second-order consequences. I agree with those who think that the Malthusian thesis of exponential growth was the capstone of Darwin's theory. "The one sentence of Malthus" acted like a crystal dropped into a supercooled fluid.

There is, however, also a second and more subtle Malthusian impact. The world of the natural theologians was an optimistic world: everything that was happening was for the common good and helped to maintain the perfect harmony of the world. The world of Malthus was a pessimistic world: there are ever-repeated catastrophes, an unending, fierce struggle for existence, yet the world essentially remains the same. However much Darwin might have begun to question the benign nature of the struggle for existence, he clearly did not appreciate the fierceness of this struggle before reading Malthus. And it permitted him to combine the best elements of Malthus and of natural theology: it brought him to the belief that the struggle for existence is not a hopeless steady-state condition, as Malthus believed, but the very means by which the harmony of the world is achieved and maintained. Adaptation is the result of the struggle for existence.

The events of September 28, 1838, are of great interest to students of theory formation. Given the extent to which Darwin was in possession of all the other pieces of his theory prior to this date, it becomes clear that in the case of a complexly structured theory it is not sufficient to have most of the pieces; one must have them all. Even a small deficiency, like defining the word "variety" typologically instead of populationally, might be sufficient to prevent the correct piecing together of the components. Equally important is the general ideological attitude of the theory-builder. A person like Edward Blyth might have had in his possession the very same components of the theory as Darwin but would have been unable to piece them together correctly owing to incompatible ideological commitments. Nothing illustrates better how important the general attitude and conceptual framework of the maker of a theory is than the simultaneous, independent proposal of the theory of natural selection by A. R. Wallace. He was one of the few people, perhaps the only one, who had had a similar set of past experiences: a life dedicated to natural history, years of collecting on tropical islands, and the experience of reading Malthus.

What Is Natural Selection?

Darwin's choice of the word "selection" was not particularly fortunate. It suggests some agent in nature who, being able to predict the future, selects "the best." This, of course, is not what natural selection does. The term simply refers to the fact that only a few (on the average, two) of all the offspring of a set of parents survive long enough to reproduce. There is no particular selective force in nature, nor a definite selecting agent. There are many possible causes for the success of the few survivors. Some survival, perhaps a lot of it, is due to stochastic processes, that is, luck. Most of it, though, is due to a superior working of the physiology of the surviving individual, which permits it to cope with the vicissitudes of the environment better than other members of the population. Selection cannot be dissected into an internal and an external portion. What determines the success of an individual is precisely the ability of the internal machinery of the organism's body (including its immune system) to cope with the challenges of the environment. It is not the environment that selects, but the organism that copes with the environment more or less successfully. There is no external selection force.

A few examples may illustrate this. Let us take, for instance, resistance against pathogens. Bacteria and other pathogens represent the environment; an animal's defense against them consists of intracellular selection processes. Likewise, adaptation to the temperature of the environment is controlled by balanced physiological mechanisms, regulated by feedback mechanisms. The success of an organism depends to a great deal on its normal development from the fertilized egg to adulthood. Almost all departures from normalcy in development will be selected against.

Considering that for many people the term "selection" has a teleological connotation—that is, it suggests purpose—many alternative terms have been proposed, such as "survival of the fittest," "selective retention," "biased nonelimination," and so forth. What all these terms try to make clear is that selection is an *a posteriori* phenomenon—that is, it is the survival of a few individuals who are either luckier than the other members of the population or who have certain attributes that give them superiority in the particular context. The probabilistic nature of selection cannot be stressed too strongly. It is not a deterministic process. Moreover, because selection is a very broad principle, it is probably not refutable (Tuomi 1981). However, each concrete application of the principle of natural selection to a specific situation is testable and refutable.

One must distinguish between two applications of selection. "Selection of" specifies the target of selection, and this is normally (in sexually reproducing organisms) a potentially reproducing individual, as represented by its phenotype (body). For this reason it is confusing to say that the gene is the unit of selection. "Selection for" specifies the particular phenotypic attribute and corresponding component of the genotype (DNA) that is responsible for the success of the selected individual. The now-obsolete concept that evolution is the interplay between genetic mutation and selection was part of the saltationist thinking of the Mendelians, as we will see. The material with which selection works is not mutation but is rather the recombination of parental genes, which produces the new genotypes that direct the development of individuals which are then exposed to selection in the next generation. It must always be remembered that selection is a two-step process. The first step consists in the production (through genetic recombination) of an immense amount of new genetic variation, while the second step is the nonrandom retention (survival) of a few of the new genetic variants.

Selection at the level of the whole organism results in changes at two other levels: that of the gene, where through the selection of individuals certain genes may increase or decrease in frequency in the population, and at the level of the species, where the selective superiority of members of one species may lead to the extinction of another species. This process, as mentioned earlier, has often been called species selection but is perhaps better called species replacement or species succession, in order to avoid misinterpretations. (Nothing is ever selected "for the good of the species.")

Finally, it must be pointed out that two kinds of qualities are at a premium in selection. What Darwin called "natural selection" refers to any attribute that favors survival, such as a better use of resources, a better adaptation to weather and climate, superior resistance to diseases, and a greater ability to escape enemies. However, an individual may make a higher genetic contribution to the next generation not by having superior survival attributes but merely by being more successful in reproduction. Darwin called this kind of selection "sexual selection." He was particularly impressed by male secondary sexual characters, such as the gorgeous plumes of male birds of paradise, the gigantic size of bull elephant seals, or the impressive antlers of stags. Modern research has shown that selection favors their evolution either because they aid in competition with other males over access to females, or because females are attracted to mates with these characteristics. This latter process is known as "female choice." Selection for reproductive success affects many life history traits beyond sexual dimorphism.

The path by which the theory of evolution by natural selection was gradually clarified and modified will be described in the following chapters. Eventually the theory was universally adopted among biologists, a development I refer to as the second Darwinian revolution.

CHAPTER SEVEN

What Is Darwinism?

C HARLES DARWIN was the most talked about person of the 1860s. T. H. Huxley, always a coiner of felicitous phrases, soon referred to Darwin's ideas as "Darwinism" (1864), and in 1889 Alfred Russel Wallace published a whole volume entitled *Darwinism*. However, since the 1860s no two authors have used the word "Darwinism" in exactly the same way. As in the old story of the three blind men and the elephant, every writer on Darwinism seemed to touch upon only one of the many aspects of Darwinism, all the while thinking that he had the real essence of what this term signifies. Thus, everybody who read the *Origin* responded only to those parts of it that either supported his own preconceived ideas or were in conflict with them. What these writers failed to grasp is that Darwinism is not a monolithic theory that rises or falls depending on the validity or invalidity of a single idea.

This monolithic tradition actually started with Darwin himself, who often spoke of his "theory of descent with modification through natural selection" (1859:459) as though the theory of common descent was inseparable from that of natural selection. How separable the two theories actually were was demonstrated when almost every knowledgeable biologist adopted the theory of common descent soon after 1859 but rejected natural selection. They explained descent instead by Lamarckian, finalist, or saltational theories (Bowler 1988). A number of passages in the *Origin* indicate how confused Darwin himself was on the subject: "The fact, as we have seen, that all past and present organic beings constitute one grand natural system, with group subordinate to group, and with extinct groups often falling inbetween recent groups, is intelligible on the theory of natural selection" (1859:478). Actually, the hierarchical organization of the living world is explicable by the theory of common descent, but this tells us absolutely nothing about the mechanism by which these changes were effected.

Even in modern times far more authors speak of Darwin's theory in the singular than acknowledge the heterogeneity of Darwin's paradigm. Even authors like Kitcher (1985) and Burian (1989), who are aware of the complexity of Darwin's paradigm, continue to refer to Darwin's theory in the singular. Burian calls the synthetic theory of evolution "the current variant of Darwin's theory."

How Darwinism is seen depends to a large extent on the background and the interests of the viewer. The word has a different meaning for a theologian, a Lamarckian, a Mendelian, or a post-synthesis evolutionary biologist. Another dimension that contributes to the diversity of opinion about the meaning of Darwinism is geography: the word "Darwinism" has meant something different in England, in Germany, in Russia, and in France. From the beginning, as we have seen, Darwin's theories were in opposition to a number of ideologies such as essentialism, physicalism, natural theology, and finalism whose strength varied from one country to the next. For the supporters of one or the other of these ideologies, the word "Darwinism" stood for the opposite of their own beliefs.

An equally great diversity exists in the time dimension. Concepts differ from facts by continuing to change over time. Hull (1985) has rightly referred to "conceptual development as a genuinely temporal process in which real change occurs." What was called Darwinism in 1859 was no longer considered so thirty years later, because the term had been transferred to something very different from that which it designated at the earlier period.

This fact was well stated by Wallace in the preface to *Darwinism* (1889). Here he explained that Darwin had done such an excellent job in proving descent with modification that this theory was now universally accepted as the order of nature in the organic world. "The objections now made to Darwin's theory apply, solely, to the particular means by which the change of species has been brought about, not to the fact of that change." Alas, Wallace was way ahead of his time in his championship of natural selection.

Different components of Darwin's paradigm were particularly interesting at different periods. At each stage in the history of Darwinism a different one of Darwin's theories was referred to as Darwinism: anticreationism vs. Christian orthodoxy, gradualism vs. Mendelian saltationism, selectionism vs. Lamarckism or finalism, and so on. This continuing change of meaning poses the awkward question of what establishes the continuity among all these Darwinisms? Do these various Darwinisms have anything in common? The answer of course is that they are all based on Darwin's original paradigm, as presented in the *Origin*.

The best way to document the great variety of meanings of the term Darwinism is to present a list of different interpretations of the term, as encountered in the literature. In each case I will attempt to analyze the validity of such usage and the temporal and ideological context in which the term Darwinism was used with this meaning. This historical analysis will then permit us to ask whether any one of the suggested definitions of Darwinism can be singled out as the best one, or perhaps even the only correct one.

Darwinism as "Darwin's Theory of Evolution"

But which one is meant, since Darwin had so many theories of evolution? Should the term refer to the totality of Darwin's theories, including those of pangenes, the effect of use and disuse, blending inheritance, and the frequency of sympatric speciation?

Surely not. To call such a conglomerate Darwinism would be worse than useless; it would be utterly misleading.

Darwinism as Evolutionism

Evolutionism was a concept alien to the physicists, not only owing to its rejection of essentialism but also for its acceptance of the historical element, so conspicuously missing from the physics of the mid-nineteenth century. Historical inferences were equally alien to all philosophers coming from logic or mathematics. It was Darwin who made evolutionary thinking a respectable concept of science. Nevertheless, it would be misleading to refer to evolutionism as Darwinism. Evolutionary thinking was already widespread when Darwin published the *Origin,* particularly in linguistics and in sociology (Toulmin 1972:326). For its existence in biology one only needs to mention the names of Buffon, Lamarck, Geoffroy, Chambers, and several German authors. Clearly, Darwin was not the father of evolutionism, even though he eventually brought about its victory.

Darwinism as Anticreationism

This is the Darwinism which denied the constancy of species and, in particular, special creation, that is, the separate creation of every feature in the inanimate and living world. There were two very different groups of anticreationists. The deists maintained a belief in God but made him a rather remote lawgiver, who did not interfere with any specific happening in this world, having already arranged for everything through his laws. Whatever happened during evolution was the result of these laws. This thought made evolution palatable to a number of Christian scientists such as Charles Lyell and Asa Gray. However, only transformational evolution—the orderly change in a lineage over time, directed toward the goal of perfect adaptation—is susceptible to this deistic interpretation. Darwin's variational evolution, with its random components at the level of both genetic recombination and selection, cannot be instrumented by strict laws. The agnostic anticreationists explained all evolutionary phenomena without invoking any supernatural agents.

Immediately after 1859 the word Darwinism simply meant a rejection of special creation. If someone rejected special creation and adopted instead the inconstancy of species, common descent, and the incorporation of man into the general evolutionary stream, he was a Darwinian. Neither natural selection nor any special theory of speciation, nor even one's belief in gradual versus saltational evolution, had any relevance to whether at that time one was considered a Darwinian or not.

There were great differences between Darwin and other "Darwinians," such as Huxley, Lyell, Wallace, and Gray, on other aspects of evolutionary theory. But these differences were of minor importance in the 1860s, because the foremost meaning of Darwinism at that time was the rejection of special creation, together with the adoption of the inconstancy of species, the theory of common descent, and (excepting Wallace) the incorporation of man into the animal kingdom. When someone in the 1860s or 1870s attacked Darwinism, he did so primarily in defense of creationism or natural theology against these four Darwinian concepts.

Perhaps the best way to determine what we should consider to be the gist of Darwinism would be to determine what Darwin had in mind when he said the *Origin* was "one long argument." Gillespie (1979), in a careful analysis of this question, concluded that it was Darwin's argument against special creation. Independently of Gillespie I had come to the same conclusion when, in the process of making a new index to the *Origin,* I saw on how many pages Darwin reiterated his conclusion that a particular phenomenon could not possibly be explained by special creation. Instead Darwin argued for a materialistic—that is, natural—explanation for the diversity of the organic world and its history. What Darwin pointed out again and again was that any given phenomenon for which special creation had been invoked could be explained much better by his theory of common descent.

Biogeography was of particular importance in that connection, because no other evidence is as convincing for common descent as the distribution of species and higher taxa. For this reason Darwin devoted two whole chapters of the *Origin* to biogeography; and in these chapters he showed over and over how a particular distribution pattern could readily be explained by common descent but not by special creation. It was the theory of common descent which had the great unifying capacity about which Darwin talked so often (Kitcher 1985:171, 184–185), because it at once gave meaning to the Linnaean hierarchy, the archetypes of the idealistic morphologists, the history of biota, and many other biological phenomena.

Curiously, many modern authors have claimed that Darwin's "one long argument" was an argument in favor of natural selection (Recker 1987). There is no good evidence for this interpretation. In his correspondence Darwin referred to his manuscript always as his "species book," not his book on natural selection. Natural selection is dealt with in the first four chapters, apparently in order to satisfy the *vera causa* demand of the leading philosophers of science (Hodge 1982); in the remaining ten chapters natural selection is not featured. Instead, these chapters are almost exclusively devoted to documentations for common descent. Indeed, Darwin himself repeatedly called attention to "the later chapters" of the *Origin* as particularly convincing proof of his theory. The facts confirm Darwin's claim. It was the reading of these chapters that converted Darwin's contemporaries to a belief in the inconstancy of species and the validity of common descent, hence in evolution. Neither these chapters nor the first four chapters, however, produced many adherents to his theory of natural selection, a theory not adopted even by some of Darwin's closest friends and followers, such as Huxley and Lyell. This is not so surprising once one realizes that the *Origin* is not one long argument in favor of natural selection.

The adoption of evolution by natural selection necessitated a complete ideological upheaval. The "hand of God" was replaced by the working of natural processes. God was "dethroned," as one of Darwin's critics formulated it. Indeed, God did not play any role in Darwin's explanatory schemes.

By what did Darwin replace him? What were the forces that played the same role in Darwin's explanations that God played in Christian dogma? Some physicalists, as well as historians and philosophers adopting physicalist explanations, suggested that Darwin had adopted Newton's reductionist explanation—that the processes of the inanimate as well as of the living world consist of "a lawbound system of matter in motion" (Greene 1987). This wording fails to reflect Darwin's thinking. The biological explanations adopted by Darwin are a long way from the Newtonian explanation, and one looks in the *Origin* in vain for any reference to "matter in motion."

Darwinism as Anti-ideology

Not only natural selection but also many other aspects of Darwin's paradigm were in complete opposition to many of the dominant ideologies of the mid-nineteenth century, as we have seen. In addition to the belief in special creation and the design argument of natural theology, other ideologies that were in total opposition to Darwin's thinking were essentialism (typology), physicalism (reductionism), and finalism (teleology). The adherents of these creeds saw in Darwin's work their most formidable opposition, and whatever the *Origin* said or implied that was dangerous to their own position was designated by them as Darwinism. But one by one these three ideologies were defeated, and with their demise the concepts of determinism, predictability, progress, and perfectability in the living world were weakened.

One of the by-products of the total refutation of all finalistic aspects of organic evolution was the inevitable reinterpretation of evolution as a historical process subject to temporary contingencies. This led to the stress on opportunism in selection and on the tinkering aspects of evolution. Such a concept of evolution is totally different from all simple transformational changes in the inanimate world, which, though also affected by chance, are primarily controlled by natural laws and thus permit rather strict predictions. A closer approach is provided by complex inanimate systems, such as the weather systems, ocean currents (greatly affected by turbulence), and the interaction of continental plates (resulting in earthquakes and volcanic eruptions), where the multitude of interacting factors and the frequency of stochastic processes defy prediction.

Darwinism as Selectionism

Almost any modern biologist, when asked what the term Darwinism stands for, will answer that it stands for a belief in the importance of natural selection in evolution. This interpretation of Darwinism is so widely accepted today that it is sometimes forgotten how relatively new this modern version is. Natural selection as the mechanism of evolutionary change was not universally adopted by biologists until the period of the evolutionary synthesis (1930s–40s); however, natural selection always was the key theory in Darwin's total research program for at least some evolutionists. The first one of whom we know that this was true was August Weismann (see Chapter 8), but he was enthusiastically followed by A. R. Wallace, who named his book on Darwinian evolution *Darwinism* because, as he said, "my work tends forceably to illustrate the overwhelming importance of natural selection over all agencies in the production of new species." However, it took another fifty years and the refutation of the major anti-Darwinian theories before this insight was generally adopted.

Darwinism as Variational Evolution

Darwin's concept of evolution was radically different from the transformational and the saltational concepts of evolution that had previously been proposed (see Chapter 4). Although it should have been obvious from the very beginning how different Darwin's concept was, this difference was not fully appreciated until very modern times. Darwin's variational evolution was based on entirely new philosophical concepts, and the neglect of these concepts is the reason why it was usually confused with the prevailing evolutionary ideas. Kitcher's discussions indicate how far even some modern philosophers are from fully understanding the magnitude of Darwin's departure from conventional thinking. Kitcher claims, "The trouble is that the theory I have ascribed to Darwin is uncontroversial—so uncontroversial as to border on triviality" (1985:30). Kitcher claims that virtually all of Darwin's opponents accepted Darwin's statement that there is variation among the members of the species and that different organisms have different properties. However, what Kitcher fails to acknowledge is that the essentialist, as was emphasized by Lyell, Sedgwick, Herschel, and others, believes that there is a definite limit to the amount of variation possible within a species or, to put it differently, permissible to a single inherent essence. For the population thinker, variation is unlimited. Hence, there is a definite possibility of going beyond the confines of an existing species. The nature and extent of variability was the crucial difference between the population thinker Darwin and his essentialistic opponents. To call this difference uncontroversial completely ignores its revolutionary importance.

Darwinism as the Creed of the Darwinians

Thwarted by the complexity of the Darwinian paradigm in their attempts to reach a satisfactory definition of Darwinism, a number of historians and philosophers have attempted in recent years to define Darwinism as the creed of the Darwinians. Hull (1985) and Kitcher (1985), in particular, have adopted this approach. This choice of defining Darwinism is more often favored by philosophers and historians than by biologists. It seems that a philosopher, particularly if he or she is coming from the background of logic or mathematics, feels on safer ground discussing a sociological phenomenon like the Darwinians than a conceptual framework like Darwinism, which requires a thorough knowledge of evolutionary biology. They attempt to justify their approach by claiming that Darwinism can best be demarcated by the scientific community (the Darwinians) who support this system. However, Recker (1990) has pointed out how questionable this claim is. Indeed, in some respects, this approach compounds the difficulties because it is as difficult to define a Darwinian as it is to define Darwinism. For Hull, "the Darwinians formed a fairly cohesive social group." Yet, this group presumably included only Lyell, Huxley, and Hooker, because Asa Gray was in America, Wallace was in the East Indies, Fritz Müller was in South America, and Haeckel was in Germany.

Recker (1990), Hull (1985), and others have stated repeatedly that there are no Darwinian tenets that characterize all the Darwinians. Indeed, it is true that some of the leading Darwinians, like Huxley and Lyell, never believed in natural selection; neither Huxley nor, presumably, Lyell endorsed Darwin's complete gradualism; neither Wallace nor Lyell thought that human beings could be dealt with in the same way as animal species. Thus, there were drastic differences among all of these Darwinians. This disparity led Hull to state that it was not sufficient for a person to hold certain Darwinian ideas in order to be called a Darwinian. In fact, he and others implied there was not a single concept subscribed to by all of the so-called Darwinians.

This is an error. There is indeed one belief that all true original Darwinians held in common, and that was their rejection of creationism, their rejection of special creation. This was the flag around which they assembled and under which they marched. When Hull claimed that "the Darwinians did not totally agree with each other, even over essentials" (1985:785), he overlooked one essential on which all these Darwinians agreed. Nothing was more essential for them than to decide whether evolution is a natural phenomenon or something controlled by God. The conviction that the diversity of the natural world was the result of natural processes and not the work of God was the idea that brought all the so-called Darwinians together in spite of their disagreements on other of Darwin's theories, and in spite of the retention by some of them (Gray, Wallace) of other theological arguments. This situation was quite well understood in the post-*Origin* period and that is why at that time, for Darwin's opponents, Darwinism simply meant denying special creation and replacing it with the theory of evolution and in particular the theory of common descent.

The theory of evolution by natural means was powerfully supported by the explanatory power of the theory of common descent. Indeed, it was this theory which eventually brought even the morphological idealists into the Darwinian camp when they realized that this was the only reasonable way to explain the hierarchical arrangement of morphological archetypes. To be sure, some morphologists, like Louis Agassiz, ascribed this order of nature to God's laws. But the natural explanation of common descent by Darwin and his followers was so much more convincing that Agassiz's interpretation fell on deaf ears and was no longer heard after Agassiz's death in 1873.

The criteria by which one delimits scientific communities must be ranked according to their importance. Creation or not was the overwhelmingly important consideration in 1859. And it was the adoption or rejection of Darwin's thesis of evolution by natural means that neatly separated the Darwinians from the non-Darwinians. They did not need to form a closely knit social group and might reside in Europe, South America, or the East Indies, but they were held together and easily recognizable by a single firm belief, that of evolution by natural means. And this explains what has puzzled some historians, that so many nineteenth-century evolutionists considered themselves to be Darwinians even though they had adopted explanatory mechanisms that were quite different from Darwin's natural selection. Only the beliefs they shared with Darwin were considered by them the truly crucial aspects of Darwinism.

To be sure, there were some borderline cases, like Asa Gray, who seems to have adopted all of Darwin's theories and yet still thought that God ultimately controlled everything, including the nature of variation that was available to natural selection. Richard

Owen was another borderline case because he really did believe in evolution of some sort, but he thought he needed to attack Darwin's theories uncompromisingly, in part owing to his enmity with Huxley.

Finally, there is the puzzling case of Lyell (Recker 1990). Even though a friend and mentor of Darwin and usually considered a Darwinian, he never accepted the most basic components of Darwin's research program. God, for him, was apparently always the ultimate causation; Lyell did not extend evolution to man, and he never accepted natural selection. If one wants to explain why Lyell continued to support Darwin even though he disagreed with most of Darwin's evolutionary beliefs, one must not forget that Darwin originally was a geologist—not only a geologist but very definitely a Lyellian geologist, who had strongly supported Lyell in his geological writings. It is quite possible that Lyell simply reciprocated, through his support of Darwin's ideas, for everything Darwin had done in his geological writings to support the Lyellian view. And, as Hodge has pointed out so convincingly, Darwin was a Lyellian when he began to occupy himself with biological problems, even though ultimately he rebelled against some of the most basic beliefs of his teacher. Thus, politically and socially, Lyell belonged to the Darwin party. Conceptually, he never was a real Darwinian.

Except for these borderline cases, the issue was absolutely clear-cut for Darwin's contemporaries. If someone believed that the origin of the diversity of life was due to natural causes, then he was a Darwinian. But if he believed that the living world was the product of creation, then he was an anti-Darwinian. The existence of a few deistic borderline cases is no more a refutation of this basic classification than the existence of a number of incipient species is an argument against the existence of species.

Darwinism as a New Worldview

J. C. Greene (1986) has suggested, in line with the view of other historians of ideas, that the affix *-ism* should be used only for ideologies and not for scientific theories. I agree with him that one should not dignify adherence to ordinary scientific theories with the ending *-ism*. However, there are scientific theories that have become important pillars of ideologies, as is the case in Newtonianism, and this is certainly true for Darwinism. Some of Darwin's important new concepts, like variational evolution, natural selection, the interplay of chance and necessity, the absence of supernatural agents in evolution, the position of man in the realm of life, and others, are not only scientific theories but are at the same time important philosophical concepts, and characterize worldviews that have incorporated these concepts. Thus, as far as several of Darwin's most basic scientific theories are concerned, they have a legitimate standing both in science and in philosophy.

The rejection of special creation signified the destruction of a previously ruling worldview. This is why not only scientists like Sedgwick and Agassiz, but also philosophers like Whewell and Herschel, opposed Darwin so vigorously. Was there a new worldview that took the place of creationism? If so, what was it and how could it be defined? Greene suggests that henceforth "the word Darwinism should be used to designate a world view that seems to have been arrived at more or less independently by Spencer, Darwin, Huxley, and Wallace, in the late 1850s and early 1860s." Here he refers to a Victorian worldview in which certain sociological ideas were used to develop a new social theory. It was based in part on the writings of Adam Smith, Malthus, and David Ricardo, and postulated that competition, struggle, and the increase in populations would result in progress. Darwin was familiar with these ideas, but as has been shown by various careful analyses of the writings of Darwin and the social philosophers, these ideas were not the source of Darwin's biological ideas (see Gordon 1989), as much as some political writers would want us to believe this.

In his writings Darwin never upheld such a worldview. Being unable to find any support for his claims in Darwin's writings, Greene suggests that we adopt Herbert Spencer's worldview, which, according to Greene, was very much the same as that of

Darwin. Spencer's worldview is described by Greene as follows: "As a lawbound system of matter in motion, evolutionary deism verging toward agnosticism under the influence of positivistic empiricism, the idea of organic evolution, the idea of a social science of historical development, faith in the beneficient effects of competitive struggle," to single out the most important points. Was this really Darwin's worldview?

There are indications that Darwin shared in the prevailing beliefs of many enlightened upperclass Englishmen, including (according to Greene) Spencer, Huxley, and Wallace. But looking critically at the list of these beliefs, as presented by Greene, one discovers that it does not include a single one that was original with Darwin, in fact, not even a single one original with Spencer. Indeed, most of these views go all the way back to the eighteenth century, even though some of them had changed their meaning, such as the term "struggle for existence," which already had been part of natural theology. Even though Greene adopts the term "Darwinism" for this set of ideas, he admits, "Spencer could rightfully demand that it be called Spencerianism," because it does not include a single one of Darwin's own ideas. What is worse, this Spencerian paradigm is in several respects in complete conflict with Darwin's ideas. For instance, Spencer supported transformational rather than variational evolution; second, his evolution was distinctly teleological; and finally, it was based entirely on an inheritance of acquired characters, not involving natural selection in any manner. Hence, it was not only scientifically, but also philosophically, something quite different from Darwin's set of ideas. To claim that Darwin and Spencer supported the same paradigm is a clear falsification of history. It is a popular thesis with sociologists, but biologists who have looked into this problem in recent years have been unanimous in refuting it (Freeman 1974).

How long did the "Darwinism," as defined by Greene, prevail as a worldview? According to Greene, through the 1860s, and perhaps up to about 1875, but Spencer, Wallace, and Huxley actually abandoned one plank or the other of this platform soon after the 1860s. The consequences of scientific Darwinism made the acceptance of this social theory of Darwinism quite untenable.

Greene suggests that certain modern biologists, like Julian Huxley, George Gaylord Simpson, and perhaps Edward O. Wilson, have an updated Darwinian worldview. The truth of the matter is that unless a person is still an adherent of creationism and believes in the literal truth of every word in the Bible, every modern thinker—any modern person who has a worldview—is in the last analysis a Darwinian. The rejection of special creation, the inclusion of man into the realm of the living world (the elimination of the special position of man versus the animals), and various other beliefs of every enlightened modern person are ultimately all based on the consequences of the theories contained in the *Origin of Species*. Nevertheless, to define Darwinism as the worldview supposedly held by Darwin in the 1860s would be about the most useless definition I can imagine.

Darwinism as a New Methodology

In view of the intense preoccupation of modern philosophy of science with questions of methodology, it is not surprising that several philosophers have asked themselves what Darwin's method of science was, and what was new in it. Was Darwin's method strictly hypothetico-deductive, as suggested by Ghiselin (1969), or did he follow various other schemes? There have been diverging answers to these questions (Recker 1987). Almost any modern philosopher of science has suggested a somewhat different Darwinian methodology. How inductive was Darwin? Does the semantic approach to the philosophy of science describe Darwin's approach best? These are the sorts of questions being asked.

Darwin realized that to convince his readers of the validity of his concepts he had to adopt a methodology that was rather different from that used by physicists to demonstrate the validity of universal physical laws. Darwin's method was to present the evidence on which he based his inferences, and he used these inferences to support his conjectures. The greater the number and the variety of pieces of evidence he could cite, the more convincing the inferences became (Ghiselin 1969). Kitcher (1985) belittled Darwin's massive citing of supporting facts, not understanding that these facts were of little interest as such but only as evidence and documentation for the inferences that Darwin was making. These inferences not only served to support his conjectures, but they also reveal some of Darwin's basic underlying ideas.

One of the reasons why different authors were able to claim that Darwin had used different methodologies is that Darwin was rather pluralistic, methodologically. In some arguments he did indeed follow the hypothetico-deductive method; in others he was proceeding in an inductive manner. I think that any claim that Darwin consistently applied only a single method could easily be refuted. And what is more important, the ultimate validation of most of Darwin's theories did not result from the victory of his methodology but from additional facts and the gradual refutation of opposing ideologies.

Many authors have called attention to the spectacular unifying capacity of Darwin's paradigm. As Dobzhansky stated it, "Nothing in biology makes sense except in the light of evolution"; I might modify this by saying "in the light of Darwinian evolution." I think it is quite misleading to suggest, as is done by Kitcher, that it was the goodness of his methodology that had this unifying effect.

There is a widespread belief among philosophers that a theory has virtually no chance of being accepted unless an appropriate mechanism is proposed simultaneously. This is indeed often true. Wegener's theory of continental drift was not accepted by the geophysicists until the mechanism for the movement of continents had been elucidated in the theory of plate tectonics. It is not *necessarily* true, however, as shown by the theory of common descent. Darwin's proposed mechanism, natural selection, was almost universally rejected, but since the fact of evolution and the theory of common descent were so completely convincing after Darwin had pointed them out, other evolutionists simply adopted—instead of natural selection—various other kinds of mechanisms, whether teleological, Lamarckian, or saltational. Indeed, for Darwin himself, as much as he believed in natural selection all his life, it was obviously not this mechanism that was of first importance for him but the evidence for evolution and common descent. Hence the completely unbalanced assignment of space to these two subjects in the *Origin*.

The version of Darwinism that developed during the evolutionary synthesis was characterized by its balanced emphasis both on natural selection and on stochastic processes; by its belief that neither evolution as a whole, nor natural selection in particular cases, is deterministic but rather that both are probabilistic processes; by its emphasis that the origin of diversity is as important a component of evolution as is adaptation; and by its realization that selection for reproductive success is as important a process in evolution as selection for survival qualities.

The Pluralism of Darwinism

It is now clear that no simple answer can be given to the question "What is Darwinism?" When someone asks this question, he is bound to receive a different answer depending on the time that has passed since 1859 and on the ideology of the person that was asked. Such pluralism is not congenial to many philosophers, and they have been trying to find some method by which they could attach the term Darwinism to a very definite meaning. The suggestion was made, for instance, that one should select an "exemplar" to fix the meaning of the term Darwinism, analogous to the action of a taxonomist who selects a type specimen in order to anchor a species to a definite name. Hull (1983; 1985) has indeed attempted to do so, but I have shown elsewhere (1983, 1989) what insurmountable difficulties oppose the application of the exemplar method.

The majority of the nine meanings of the term Darwinism discussed above are clearly either misleading or unrepresentative of Darwin's thought. Looking at the situation as a historian, I am impressed that two meanings have had the widest acceptance. After 1859, that is, during the first Darwinian revolution, Darwinism for almost everybody meant explaining the living world by natural processes. As we will see, during and after the evolutionary synthesis the term "Darwinism" unanimously meant adaptive evolutionary change under the influence of natural selection, and variational instead of transformational evolution. These are the only two truly meaningful concepts of Darwinism, the one ruling in the nineteenth century (and up to about 1930), the other ruling in the twentieth century (a consensus having been reached during the evolutionary synthesis). Any other use of the term Darwinism by a modern author is bound to be misleading.

CHAPTER EIGHT

A Hard Look at Soft Inheritance: Neo-Darwinism

DARWIN HIMSELF was in part responsible for the difficulties his theories encountered in the years following the publication of the *Origin*. On most any subject he dealt with—and this includes nearly all of his theories—he not infrequently reversed himself. One of Darwin's twentieth-century critics was not entirely unjust in asserting that "Darwin's hedging and self contradiction enable any unscrupulous reader to choose his text from the *Origin of Species* or *The Descent of Man* with almost the same ease or accommodation to his purpose as if he had chosen from the Bible" (Barzun 1958:75). For instance, although in principle Darwin rejected essentialism and explained adaptation in strictly variational terms (the individual was for him the target of selection), he sometimes fell back into typological language, as for the origin of an incipient species by the isolation of a variety. Darwin strongly rejected any teleological "law of necessary development" (1859:351), and yet he made a sufficient number of careless statements to permit certain historians to consider Darwin a teleologist (Himmelfarb 1959). Although he often stated that enough variability was available for natural selection to explain the origin of every adaptation, he nevertheless declared his puzzlement over the evolution of eyes. On one page he might say that only slight variations are evolutionarily important, but then on another page he will discuss rather strikingly different varieties, like the ancon sheep and the turnspit dog, both of them extremely short-legged. Kohn (1989) has made a detailed analysis of Darwin's vacillation and its psychological and tactical reasons.

It is frequently stated that Darwin totally rejected all Lamarckian ideas in the first edition of the *Origin* and that he allowed at that time no mechanisms of evolution other than random variation and natural selection. This is not correct. Darwin displayed considerable indecision already in 1859 on the origin of variation and the nature of inheritance. He makes no less than three sets of concessions to the possibility that the environment in the widest sense of the word can induce genetic variation and that these acquired characters can be inherited. First, he speculated about a direct effect of the environment on certain structures; second, he hypothesized an indirect effect of the environment in increasing variability; and third, he discussed the effects of use and disuse, for instance, when he says that the reduced size of the eyes of moles and other burrowing mammals is "probably due to gradual reduction from disuse but aided perhaps by natural selection" (1859:137). In the case of cave animals, when speaking of the loss of eyes, he says, "I attribute their loss wholly to disuse." On the other hand, he calls attention to the fact that the adaptations of the neutral castes of social insects are inexplicable with any kind of Lamarckian theory.

Darwin's major thesis was that evolutionary change is due to the production of variation in a population and the survival and reproductive success ("selection") of some of these variants. But the origin of this variation puzzled him all his life. Darwin considered variation to be an intermittent phenomenon, one that occurred mostly under special circumstances. However, he was quite sure that in nature there was an immense reservoir of variation that was always available as the material of selection. Darwin devoted the entire fifth chapter of the *Origin* to the "laws of variation" and followed it up some years later with the two-volume work *The Variation of Animals and Plants under Domestication* (1868). Two years earlier, unknown to Darwin, Gregor Mendel had published the key to an understanding of the nature of variation. But Darwin never heard of Mendel's work and was never able to solve the problem. Under the influence of the prevailing concepts of physicalism, Darwin looked for a direct cause of each variant. Also, his partial belief in blending inheritance—the fusion of the inheritable contribution of the parents in the off-spring—made it impossible for Darwin to solve the problem. His partly typological, partly deterministic approach precluded his finding the right answers.

The great number of misconceptions about the nature of inheritance and the absence of good empirical data led one of Darwin's younger contemporaries to conclude that the study of variation was the new frontier of evolutionary biology. This young evolutionist was August Weismann (1834–1914). By his uncompromising refutation of an inheritance of acquired characters, and by his genetical theorizing, even when wrong, he laid the foundation for the acceptance of Mendelian inheritance, the next major advance toward modern evolutionism.

In the first twenty years after the publication of the *Origin* no conceptual contributions were made to the Darwinian paradigm. That is, none were having any impact. To be sure, Mendel published in 1866 his fundamental discoveries about inheritance, but they were completely ignored (until 1900) and left contemporary evolutionary thinking unaffected. A decisive new departure did not happen until 1883, when Weismann published his provocative essay *On Heredity* (1883). Here, Weismann denied categorically any occurrence of the effects of use or disuse, indeed any inheritance of acquired characters. In this uncompromising position he was immediately joined by A. R. Wallace, who pleaded for the monopoly of selection in his book *Darwinism* (1889). George J. Romanes, a disciple in Darwin's later years, considered the denial of soft inheritance a major departure from Darwin's own teaching and therefore in 1896 coined the term "neo-Darwinism" for Weismann's exclusive selectionism, or, as Weismann himself once put it, Darwinism without inheritance of acquired characters. This term continued to be used through the evolutionary synthesis, but once a belief in soft inheritance had been universally abandoned, most people reverted to the simple term "Darwinism." In the years between 1883 and the evolutionary synthesis, the fight against soft inheritance was intimately connected with the name Weismann.

August Weismann is one of the towering figures in the history of evolutionary biology. If we ask who in the nineteenth century after Darwin had the greatest impact on evolutionary theory, the unequivocal answer must be Weismann. What makes Weismann's concern with the problems of evolution so fascinating is the gradual maturation of his thinking. There was an unfortunate tendency in the older histories of science to depict the thought of a great scientist as a monolithic structure, something poured out of one mould, impressively unchanging. More recent researches have shown how misleading such a presentation is, and this is true for almost any scientist one may mention. When Weismann rejected the theory of an inheritance of acquired characters, surely a dramatic conversion, he was already over forty-seven years of age. This change required a fundamental rethinking of all his previous assumptions on inheritance and evolution.

It may be helpful to recognize three periods in Weismann's thinking: 1868–1881 or 1882, during which he accepted the inheritance of acquired characters; 1882–1895, when he searched for a source of genetic variation; and 1896–1910, when he recognized germinal selection as the aid of natural selection.

In 1859, when Darwin published his revolutionary *On the Origin of Species,* Weismann was twenty-five years old. The subject of evolution was never raised during his student days at several German universities. Nor did Weismann's early zoological work have anything to do with evolution; it dealt instead with such topics as muscle histology and insect embryology. And yet, in 1868, Weismann chose Darwinism as the subject of his inaugural lecture at Freiburg. In no other country, not even in England, did Darwinism have the impact it had in Germany, as is evident from the writings of contemporary zoologists, botanists, and anatomists. The high importance Weismann attributed to Darwin's thought is documented by the fact that he compared the "transmutation" theory to the Copernican heliocentric theory and implied that no advance in human understanding since the acceptance of that theory had had as great an impact as Darwin's theory (1868:30).

Weismann understood from the beginning that one must distinguish two theories of evolution; evolution as such, called by Weismann the transmutation theory; and Darwin's explanatory theory, the theory of natural selection.

Weismann's attitude toward evolution as such was close to that of the modern evolutionist, for whom evolution is not a theory but an accepted fact. The various conclusions arrived at by evolutionary biology, said Weismann, "may be maintained with the same degree of certainty as that with which astronomy asserts that the earth moves around the sun; for a conclusion may be arrived at as safely by other methods as by mathematical calculation" (1886:255). The fact of evolution was so irrefutably established for Weismann that in his subsequent writings he virtually never bothered to list facts in support of evolution but instead concentrated on the causal aspects of the evolutionary process.

Weismann enthusiastically accepted the theory of common descent even though the construction of phylogenies was not one of his fields of specialization. In particular he fully endorsed the theory of recapitulation and based his analysis of the ontogenetic stages of sphingid caterpillars entirely on this principle. Weismann was aware of the relative independence of evolutionary developments in the ontogeny of different orders of insects, by calling attention to the overall uniformity in the morphology of adult wasps and flies and by interpreting the numerous specializations found in the larval stages as secondary adaptations. (For a further discussion of Weismann's views on recapitulation see Gould 1977:102–109.)

Weismann's Assault on Antiselectionism

By far the most important of Darwin's evolutionary theories was natural selection. It attempted to explain with the help of material causes what had been explained previously by a supernatural cause, by design. This theory was so novel, so daring, that at first it was adopted by very few biologists. Weismann was one of the few; and, as we shall see, eventually he went even further than

Darwin in asserting the "allsufficiency" of natural selection. Most evolutionists in the post-Darwinian period supported other theories. The three most widely accepted of these opposing theories (Bowler 1983) were (1) a belief in an intrinsic driving force, or "phyletic force," resulting in evolution by "orthogenesis"; (2) saltational evolution; (3) Lamarckian factors (inheritance of acquired characters; see below).

In the 1860s and 1870s the idea of some teleological, finalistic, phyletic force was perhaps the most popular of the three options, at least in Germany. In view of the fact that it had been adopted by some of Germany's most admired biologists, such as von Baer, Naegeli, and Kölliker, by the orthogenisists Haacke and Eimer, and by leading philosophers, particularly von Hartmann, Weismann found it necessary to emphasize again and again his resistance to the acceptance of such a metaphysical force, and he attempted to refute it by ever-new arguments. In view of the prestige of his opponents, he chose not to ridicule this school of thought but rather to show that in his opinion it was inconsistent with the known facts. "How could such intrinsic drives produce such sophisticated patterns as found in leaf imitating butterflies, like Kallima?" he asked. Furthermore, if there were such an evolutionary force, he said, it should be possible to establish it by empirical research. But no one had succeeded in doing so; instead, he said, it was possible to trace all evolutionary changes "to known transforming factors," that is, to natural selection (1882:161).

The second of the great antiselectionist theories of the post-Darwinian period was the origin of new types and organs by major saltations. This contention was diametrically opposed to evolutionary gradualism, one of Darwin's five evolutionary theories. Yet saltationism was favored by T. H. Huxley, Kölliker, and other Darwin contemporaries. Weismann was unalterably opposed to the possibility of saltational evolutionary change (1882:697). It is interesting that in 1882 and 1886 Weismann already anticipated the claims made a few years later by Bateson (1894) and de Vries (1901) and attempted to refute them. "An abrupt transformation of a species is inconceivable, because it would render the species incapable of existence" (1886:264), he wrote, believing that the existence of numerous coadaptations would make such an instantaneous total restructuring of the organism an impossibility. After all, when studying organisms one is amazed at the precision and ubiquity of adaptations. "It is evident that one can not possibly explain these innumerable adaptations by the occurrence of a rare, accidentally at one time occurring, variation. The necessary variations with the help of which selection pieces together evolutionary change must always be offered again and again in many individuals" (1892:568). At the height of the popularity of de Vries's mutation theory, Weismann once more demonstrated the utter improbability of saltational evolution (1909:22–24).

Weismann was proud to be able to assert, "more definitely than Darwin has done," that all changes must have occurred "very gradually and by the smallest steps" (1886:264). It is obvious why Weismann had to insist on *gradual* evolutionary change when one remembers that at first he attributed much of it to use and disuse, and later considered evolutionary change as more or less due to quantitative changes in large numbers of determinants. Even after de Vries had published his mutation theory, Weismann postulated that a saltation of the phenotype was simply the making visible of a long series of preceding small genetic changes (1904, 2:119).

Weismann's Evidence for Selection

Rejecting all other possible causations for evolutionary change, Weismann firmly adopted selectionism. He showed himself a convinced selectionist in his very first evolutionary publication (1868). He supported his selectionism by various sets of arguments. In view of the evident invalidity of all teleological and saltational theories of evolution, said Weismann, there is really no other option than to accept natural selection. There is not a detail in the structure or physiology of an organism that has not been shaped by natural selection. Without a doubt, Weismann was the most consistent selectionist in the nineteenth century.

Natural selection was the main topic of the series of papers Weismann published in the 1870s and 1880s dealing with the theory of descent. In his detailed studies of the markings of sphingid caterpillars, he unequivocally adopted the adaptationist program. He asked: "Have the markings of the caterpillars any biological value, or are they in a measure only sports of nature? Can they be considered as partially or entirely the result of natural selection? Or has this agency had no share in their production?" (1882:308). To be able to answer these questions, Weismann developed a rigorous methodology. And he tested his assumptions with carefully executed experiments in which caterpillars that had been placed on a variety of backgrounds were exposed to birds or lizards. Furthermore, Weismann always sought to determine whether or not there was a correlation between structural or color characters and behavioral characteristics. The result of these studies was that he found abundant evidence in favor of selection. "It has been possible to show that each of the three chief elements in the markings of the Sphingidae have a biological significance, and their origin by means of natural selection has thus been made to appear probable" (1882:380).

Wherever Weismann looked, he found evidence for selection—not only in the animal kingdom but also among plants. He examined the adaptive significance of the form and color of flowers, so beautifully demonstrated by Darwin for orchids and by numerous other students of flower biology beginning with Sprengel; the venation of leaves; and the numerous protective devices of plants against herbivores. Weismann presented a superb description of the aquatic adaptations of marine mammals, demonstrating that all the differences between them and terrestrial mammals are clearly adaptations (1886:261).

Natural selection is active not only in the acquisition of new adaptations but in the maintenance of existing ones. As soon as the selection pressure is relaxed, as in the case of the eyes of cave animals, individuals with imperfect structures will no longer be eliminated. "In my opinion every organ is kept at the peak of its conformation only by continuous selection. And it slides down from this height ceaselessly even though very slowly, as soon as it no longer is of value for the survival of the species" (1893:51). Such a loss of organs can result from either a relaxation of the maintenance selection or from an actual counter-selection exercised by competition for tissue substrate. Weismann presented a particularly impressive documentation of natural selection in his *Vorträge* (1904 I; 36–170).

Weismann professed without hesitation that he was a panselectionist. "There is no part of the body of an individual or of any of its ancestors, not even the minutest and most insignificant part, which has arisen in any other way than under the influence of the conditions of life" (1886:260). He admitted, however, that "these are indeed only convictions, not real proofs." In the 1880s direct experimental proof for the force of selection was scanty. The analogy with artificial selection was still perhaps the most convincing evidence for selection; for, as Weismann said, the breeders had achieved almost any objective of their selection (1893:60). Otherwise the principle of exclusion is the best support for natural selection, a principle which states that a theory can be adopted for the time being if all competing theories have been refuted.

After Weismann had become an ardent selectionist, he realized (as had Darwin before him) that selection does not necessarily lead to perfection. "No device in nature is absolutely perfect, not even that beautifully constructed eye of man. Everything is only as perfect as necessary, at least as perfect as it must be in order to accomplish what it is supposed to accomplish." This echoes similar statements made by Darwin (1859:201, 206).

In recent years several authors have proudly announced the discovery of constraints on the action of natural selection. Weismann was aware of the role of developmental constraints more than a hundred years ago, a sign of his deep understanding of the evolutionary process. The physical constitution of each species limits the action of natural selection because it "restrains the course of development, however wide the latter may be" (1882:113–114). These developmental constraints are the reason "why organic evolution has frequently proceeded for longer or shorter periods along certain developmental lines" (1886:258).

The question of what the target of selection is occupied Weismann all his life. At first he followed Darwin, recognizing only a single level of selection, and evidently considered the individual as a whole to be the target. He considered the genotype as a holistic system, the individual components of which could not be randomly interchanged or replaced. To what extent Weismann gave up his holistic view after he had developed his complex theories of the structure of the genotype, and his postulate of three levels of selection, has not yet been determined.

As for sexual selection, Weismann at first was quite enthusiastic, and in 1872 he attributed many differences between the sexes, even in insects, to sexual selection (1872:60); indeed he felt the theory should be applied even more extensively. Weismann saw clearly that sexual selection gave an advantage to certain individuals, and he remarked that the driving force of sexual selection is not the external environment but rather the preferences of the individuals in their selection of a mate. Characters acquired by sexual selection do not offer any advantage in the daily struggle for existence (1872:62).

Ten years later Weismann (1882:101–102) had become rather confused about the cause of sexual dimorphism. Twenty years later Weismann returned completely to Darwin's position. He devoted an entire chapter of his *Vorträge* (1904 chap. 2) to sexual selection, definitely accepting the principle of female choice and attempting to refute all of Wallace's objections.

The subsequent history of evaluation of sexual selection has been one of ups and downs. When the mathematical population geneticists declared the individual gene to be the unit of selection and defined fitness as the contribution of such genes to the gene pool of the next generation, there was little room left for sexual selection. This attitude characterized the period from the 1920s to the 1960s. In the past two decades the importance of sexual selection has again been acknowledged; it has been recognized that the individual as a whole is the principal target of selection and that such an individual might have a reproductive advantage owing to characteristics that do not contribute to general fitness. This topic has become one of the major concerns of sociobiology.

Rejection of the Inheritance of Acquired Characters

When Weismann concluded in 1882 that the theory of an inheritance of acquired characters was untenable, he had to search for a new source of genetic variation and therefore devoted the second research period mostly to genetic studies. Indeed, he stated quite explicitly that his theory of inheritance "was so to speak only a means toward a higher purpose, [to establish] a foundation for the understanding of the transformation of organisms in the course of time" (1904:iii). Weismann understood from the very beginning that variation was an indispensable prerequisite for the operation of natural selection. Variation was no problem for him, because at that time he believed in "the origination of transformations by the direct action of external conditions of life" (1882:682). There is probably no other evolutionist whom historians traditionally have considered as extreme a selectionist as Weismann. So it will come as a great surprise when they learn that, at least until 1881, Weismann, like Darwin, believed in an inheritance of acquired characters. Not only did he believe in the heritable effect of use and disuse, but he stated as late as November 1881 (1882:xvii): "Nor can the transforming influence of direct action, as upheld by Lamarck, be called in question, although its extent can not as yet be estimated with any certainty." (Weismann's conversion to "hard" inheritance must have occurred between 1881 and 1883. He still supported an inheritance of acquired characters in at least three statements in the preface, dated November 1881, to the English edition of his *Studien zur Descendenztheorie*, 1876, published in 1882. In this edition Weismann added a number of footnotes on various subjects but made no disclaimers to some of the strongly Lamarckian statements of the original German text.) That variation was an indispensable component of the process of natural selection remained Weismann's firm conviction even after he had changed his mind completely about the causation of such variation. Even in 1893, he still referred to variation "as one of the main factors of natural selection" (1893:54).

The inheritance of acquired characters was for Weismann the major source of variation in all of his evolutionary writings prior to 1882. He later admitted, "Twenty-five years ago I was still of the opinion that in addition to [other factors] also the inherited effect of use and disuse played a not unimportant role in evolution" (1893:43). Actually, in Weismann's pre-1883 writings one can find numerous suggestions about how the environment might influence the variation and inheritance of organisms (Churchill 1968; Blacher 1982). What Weismann envisioned was that species and populations went through occasional periods of greatly increased variability, and that if segments of a population were isolated during such a period, then a process he called "amixia" would lead to differences. Some of Weismann's descriptions sound like what later was called "genetic drift."

Because this is often misunderstood, let me emphasize that there was no real conflict between a belief in selection and a belief in an inheritance of acquired characters—for either Weismann or Darwin. Both believed from the beginning in the overwhelming importance of natural selection as the mechanism responsible for the production of adaptation, but they desperately needed to find some mechanism that would produce the variation needed for the operation of natural selection. It was the task of the inheritance of acquired characters to supply at least part of this variation. That both authors made use of this factor was understandable, for a belief in an inheritance of acquired characters was virtually universal until the beginning of the 1880s. This was as true for folklore as for science. The few dissenters (Galton, perhaps His) spoke with muted voices and were not heard.

In view of Weismann's repeated Lamarckian statements in 1881, it is rather surprising how sweeping his repudiation of Lamarckism was in his 1883 lecture "On Heredity." His refutation of Lamarckian claims is so broad, and his Darwinian interpretation of the numerous cases that had previously been cited as proofs for an inheritance of acquired characters is so well thought out, that one is almost forced to the conclusion that Weismann must have been thinking about this problem for many years. Despite its title, Weismann's 1883 paper did not develop a theory of inheritance but was devoted almost exclusively to the refutation of an inheritance of acquired characters. His strategy was remarkably similar to Darwin's when he refuted creationism: Weismann took up one case after another that simply could not be explained by "use and disuse" and other Lamarckian mechanisms. How can the numerous special adaptations of the worker and soldier castes of ants be inherited by use, when these castes do not reproduce? How can habits become instincts through use, when a particular instinct is practiced only once in the whole life of the individual, as is so often the case of reproductive instincts among insects? How can the external structure of insects be modified by use and disuse, when the chitinous skeleton is laid down during the pupal stage and never changes afterward? To a modern person, fully convinced of the impossibility of an inheritance of acquired characters, Weismann's arguments seem most persuasive. But in Weismann's time the belief in the Lamarckian principle was so deeply ingrained that only a minority were converted. Use and disuse seemed a far more convincing explanation of the loss of extremities by snakes or of eyes by cave animals. It was not until the evolutionary synthesis of the 1940s that unreserved selectionism was more or less universally adopted by biologists; but the conclusive refutation of the principle of the inheritance of acquired characters was not achieved until the 1950s, through the so-called central dogma of molecular biology, which stated that no information contained in the properties of the somatic proteins could be transferred to the nucleic acids of DNA.

Weismann's strategy was to show not only that an inheritance of acquired characters encounters formidable difficulties but also that cases cited in its favor could be explained quite well through the theory of natural selection. A structure that is used a great deal in an individual's lifetime is of course also exposed to strong selection forces. If the organ is inferior in a particular individual, its owner will be handicapped in the struggle for existence. Therefore, "the improvement of an organ in the course of generations is not the result of a summation of the result of practice of individual lives, but of the summation of favorable genetic factors" (1883:26).

Weismann stated repeatedly that he was inclined to apply the principle of selection far more consistently than Darwin himself. The fact that in ducks and other domestic fowl the wings have somewhat degenerated while the legs have become stronger than in their wild ancestors was explained by Darwin as the result of use and disuse. Weismann quite rightly pointed out that this can be explained even better by assuming that natural selection was the cause of this change in proportions.

It is rather ironical that natural selection is attacked in the current evolutionary literature not so much for its inability to explain certain adaptations or other evolutionary developments as for being a principle so successful that anything could be explained by natural selection—and therefore, to use Karl Popper's language, that it would be impossible ever to refute any evolutionary explanation based on the principle of natural selection. This was not the situation in Weismann's day, when people had not yet become accustomed to thinking in terms of natural selection and when they were far more comfortable explaining evolutionary developments in terms of an inheritance of acquired characters.

Weismann was in his mid-forties when he shifted from Lamarckism to an uncompromising selectionism. An important role in his conversion was undoubtedly played by his own observation—and that of various cytologists and embryologists—that the future germ cells in various types of invertebrates are set aside after the first mitotic divisions of the developing embryo and no longer have any physiological connection with the body cells (Churchill 1985).

This observation led Weismann in 1885 to his theory of the "continuity of the germ plasm," which states that the "germ track" is separate from the body (soma) track from the very beginning, and thus nothing that happens to the soma can be communicated to the germ cells and their nuclei. In this early theory the separation was between germ cells and body cells. Even though such a separation may occur in certain organisms during the earliest cell divisions of development, in most organisms (particularly in plants) many if not most somatic tissues are able to produce germ cells. For this reason Weismann replaced germ cells by germ plasm in his later publications. The strict separation we now make of the DNA program of the nucleus from the proteins in the cytoplasm of each cell reflects Weismann's early insight (Churchill 1985).

Weismann was correct in concluding that the germinal material is something entirely different from the body substance and "that the differentiation of the body cells is not acquired by them directly, but that it was prepared by changes in the molecular structure of the germ cell" (1883:14).

Weismann came close to the concept of a genetic program when he said that we will come to the right conclusions about development "if we consider all processes of differentiation that occur in the course of ontogeny as controlled by the chemical and physical molecular structure of the germ cell" (1883:18).

The Significance of Sex and Genetic Recombination

The refutation of an inheritance of acquired characters seemingly left a serious void in evolutionary theory. Weismann fully understood that he had to find a new mechanism for the production of genetic variability. His solution was this: "I believe that such a source is to be looked for in . . . sexual reproduction . . . Two groups of hereditary tendencies are, as it were, combined. I regard this combination as the cause of hereditary individual characters, and I believe that the production of such characters is the true significance of [sexual] reproduction" (1886:272). Indeed, this process of genetic recombination through sexual reproduction was recognized by Weismann as being one of the most important processes in evolution, and to it he devoted an entire long essay (1886).

His conclusion was in total opposition to prevailing ideas. Since blending inheritance was widely accepted in 1880, even by Darwin (Mayr 1982:779–781), who simultaneously believed in particulate inheritance, sexual reproduction was credited with assuring the uniformity of species. That Weismann could put forth such revolutionary claims was made possible by the discovery of van Beneden (1883) and other cytologists that maternal and paternal chromosomes do not fuse during fertilization but merely reestablish the diploidy of the zygote. What sexual reproduction thus achieves is not a homogenization of the parental characters but their recombination. Genetic recombination together with natural selection can thus bring together previously separate and independent characteristics that greatly improve the selection value of their bearers. Concerning the origin of sexuality, Weismann's statements are vague, if not teleological.

To a remarkable extent Weismann was aware also of the drawbacks of sexuality. He describes the distinct advantage by which the temporary abandonment of sexual reproduction allows certain animals such as aphids and cladocerans "a much more rapid increase in the number of individuals . . . in a given time" (1886:289). However, this is only a temporary advantage—a conclusion confirmed, according to Weismann, by the fact that "whole groups of purely parthenogenetic species or genera are never met with" (1886:290). Although this statement is no longer literally true, we know of only a single higher taxon of animals, the bdelloid rotifers, in which all species are asexual. All other asexually reproducing groups of animals seem to become extinct sooner or later.

The overwhelming importance of genetic recombination was almost totally ignored by most Mendelians, owing to their strongly reductionist position. The broad recognition of recombination in the evolutionary process had to await the evolutionary synthesis. Those mathematical population geneticists who emphasized the gene as the target of selection were particularly slow to appreciate fully the role of recombination. It is today quite evident that what is of particular importance in evolution is not so much allelic interactions as interactions among different loci and different chromosomes, which are constantly changed by genetic recombination. Weismann's (1891) championship of amphimixis, as he called it, is one of his most important contributions to evolutionary biology. Weismann's explanation of the evolutionary significance of sexual reproduction has been challenged in recent years by a number of authors, but no consensus has yet been reached as to a possible alternative explanation.

The Source of New Genetic Variation

The mixing of the genetic factors of both parents produces an almost unlimited supply of genetically new individuals in every generation, but this process consists only in the intermingling of already existing variations. The origin of entirely new genetic factors remains unexplained. The origin of true genetic novelty became a crucial problem for anyone rejecting an inheritance of acquired characters, that is, a transfer of new somatic characters to the germ plasm. Weismann fully realized the seriousness of this problem and struggled with it from 1882 until far into the 1890s. He repeatedly stated that organisms normally give rise "only to exact copies of themselves" (1882:679, 682). Furthermore, natural selection is bound eventually to exhaust the supply of genetic variants available through recombination. Thus, Weismann was forced to come up with a new solution; and for this, he was ill equipped (Mayr, 1988).

A mechanistic conception of causation completely dominated Weismann's thinking about the source of genetic variation. When, prior to 1883, Weismann gave up his belief in an inheritance of acquired characters, he had to find an entirely new explanation for the origin of genetic variation. Being involved at that time in a struggle with Naegeli, Kölliker, Eimer, and others about the existence of internal forces of evolution, Weismann was unable to adopt a process corresponding to what we now call "spontaneous mutation." Would not such a process have to be caused by internal forces? This would open the door to what he called "metaphysical principles." He wanted to explain everything "mechanically," and his concept of mechanical was rather classical, implying the direct action of visible forces such as climate, nutrition, and the like. Instead, he postulated a number of other processes in the germ line, none of which turned out to be correct. Ultimately all these explanations were based on the idea of multiple replicas of genetic determinants in the germ plasm and on quantitative shifts in the relative number of these various elements.

The vigorous claims of Herbert Spencer and other neo-Lamarckians that random variation was insufficient to supply the needed material for the exercise of natural selection left their mark on Weismann's thinking. He eventually agreed with the Lamarckians that a simple Darwinian selection of individuals is not sufficient to explain all the phenomena of evolution (1896:59), for instance, the continuing reduction of vestigial organs (1896:24). Finally, Weismann admitted that chance alone could not produce the right variation in the right species at the right time. When perfect adaptation is achieved, "chance is out of the question. The variations which are supplied to the natural selection of individuals must have been produced [within the germ] by the principle of the survival of the fittest" (1896:46). And this led Weismann to propose his hypothesis of germinal selection.

This is not the place to analyze the rationale of Weismann's theory of inheritance, but one of its aspects must be mentioned because it vitally affected Weismann's evolutionary thinking. As part of the dominating influence of physicalism in nineteenth-century science, quality was considered an unscientific concept. Seemingly qualitative differences occurring in evolution had to be converted to quantitative differences. And this is precisely what Weismann believed his theory of inheritance could do. It permitted him to show that "all variation is in the last analysis quantitative, consisting of an increase or decrease of the living particles or of their constituents, the molecules" (1904 2:128). These changes do not occur spontaneously but are always caused by external factors, either by the differential nutrition of various determinants within the germ plasm or by the environment.

Weismann apparently never realized that, notwithstanding the terminology, his was no longer a selection theory. Indeed, it approached Geoffroyism, direct germinal induction by the environment, rather dangerously. Even though Weismann continued to deny emphatically any ability of the soma (phenotype) to affect the germ plasm, in his theory of induced germinal selection he admitted a direct effect of the environment on the germ plasm.

The most fundamental component of any theory of inheritance is insistence on the basic constancy of the genetic material. Yet the theory of germinal selection abandoned a constancy of the underlying genetic elements. It seems that Weismann was forced into this change of mind by the claims of various Lamarckians that exposing the pupae of certain Lepidoptera to heat or cold shocks had not only changed the coloration of the emerging butterfly but also that of some of its untreated descendants (1904 2:230–231). These results are now known to have rested on faulty experimentation, and it is an irony of fate that they caused Weismann to accept induced germinal selection and thus needlessly undermined the consistency of his theory of inheritance. It is worth noting that Darwin proposed his ill-fated theory of pangenesis also quite needlessly, because it was expressly proposed to explain the effects of use and disuse, which one year after Darwin's death were shown to be illusory (Weismann 1883).

Weismann's theory of germinal selection was full of internal contradictions and was almost unanimously rejected by his contemporaries, although he pleaded for it once more as late as 1909 (pp. 36–37), as well as in the third edition of his *Vorträge* (1910). All possible support for it had by that time been swept away by the acceptance of Mendelian inheritance. This included major conceptual shifts, among them the acceptance of spontaneous mutations and of qualitative genetic changes.

Weismann retained an unshaken faith in Darwinian natural selection all his life, in spite of the vicious attacks of Driesch and
other experimental zoologists. When the fiftieth anniversary of
the publication of the *Origin of Species* was celebrated in 1909,
Weismann published a strong manifesto on the power of natural
selection. Such a declaration required considerable faith and courage, since natural selection at that time was at the lowest point of
its scientific acceptance, owing to the attacks of de Vries, Bateson, Johannsen, and other Mendelians.

The Heritage of Weismann's Ideas

There can be little doubt that, after Darwin, no biologist in the
nineteenth century had as great an impact on evolutionary thinking as Weismann. A combination of characteristics enabled Weismann to exercise this role. He had, on the one hand, a powerful
analytic ability; he could build a logical argument step by step.
On the other hand, like Darwin, he had an extraordinary facility
for constructing hypotheses. This was by no means appreciated
in the nineteenth century; and, like Darwin, Weismann was ridiculed for his speculations. In view of the paucity of available facts
on inheritance, some of his speculations would be considered
rather daring even by a modern biologist who is constantly encouraged to try his hand at model building. When one reads the
attacks by Oscar Hertwig, Wolff, Driesch, and others, one realizes how outraged his contemporaries were by Weismann's theorizing.

What Weismann did, he did for good reason. He was aware of
the intellectual poverty of the inductionism then dominant in
Germany. "The time in which men believed that science could be
advanced by the mere collection of facts has long passed away"
(1886:295). Surely, it had passed away for Weismann, but most of
his contemporaries had not yet perceived the message. Until Darwin, Haeckel, Weismann, and a few other theorizers had begun
to develop a conceptual framework of biology, "the investigation
of mere details had led to a state of intellectual shortsightedness,

interest being shown only for that which was immediately in view. Immense numbers of detailed facts were thus accumulated, but . . . the intellectual bond which should have bound them together was wanting" (1882:xv). Weismann again and again emphasized that his hypotheses were not the ultimate truth but were proposed as heuristic devices. Of one of his theories he said, "Even if it should be later necessary to abandon this theory, nevertheless it seems to me to be a necessary stepping-stone in the development of our understanding, it was absolutely necessary that it had to be proposed, and it must be carefully analyzed, regardless of whether in the future it will be found to be correct or false" (1885:17). This is a clear statement of his hypothetico-deductive scientific philosophy.

To sum up, then, Weismann's major contributions to biological thought were:

(1) *Defense of natural selection.* From about 1890 to 1910, Darwin's theory was threatened to such an extent by various opposing theories that it was in danger of going under. Weismann's undeviating support of natural selection at this time was a major contribution to the emergence of a strengthened Darwinism. Weismann helped to develop a methodology for the analysis of natural selection, showing that it would permit the making of predictions that would be confirmed if natural selection did indeed operate. He showed that the reduction or loss of structures, which greatly disturbed some of his contemporaries, could be explained as being caused by a relaxation of selection pressure.

Weismann was one of the first to test selection and environmental influences by experiment. For instance, he exposed caterpillars of different colorations to potential predators on differently colored substrates. In other experiments he tested the effect on the color of butterflies of the temperature at which the pupa was kept.

Perhaps even more effective than his evidence in favor of selectionism were his arguments against orthogenesis, saltationism, and Lamarckism, the three theories competing with selection (Bowler 1983). He was particularly convincing in his argument in favor of the gradual nature of evolutionary change.

(2) *Refutation of the theory of the inheritance of acquired characters.* Through his sweeping rejection of an inheritance of acquired characters Weismann established a new version of Darwinism. The inheritance of acquired characters never regained full credibility after Weismann's attack in 1883. Weismann supported his case by three lines of evidence: there is no cytological mechanism that could effect such a transfer from soma to germ plasm; there are many adaptations that could not have been acquired by such an inheritance (for example, the soldier caste of ants and termites); and all reputed cases of inheritance of acquired characters can be explained by selection. Even though Weismann was occasionally wrong in detail, he was right in principle, and the basic Weismannian thought is now articulated in the so-called central dogma of molecular biology.

(3) *Firm establishment of particulate inheritance.* If there were blending inheritance (at the gene level), very different laws of genetics (Galton) would be valid than if inheritance were particulate. Furthermore, if each act of fertilization (zygote formation) consists in the combining (rather than fusion) of the paternal and maternal genomes, then this must be compensated by a reduction division, Weismann hypothesized (Churchill 1968; 1979; Farley 1982). These postulates laid the foundation for Mendelian genetics (as clearly stated by Correns), and Mendelian genetics in its turn validated Weismann's theories.

(4) *Recognition of the importance of sexual reproduction as a source of genetic variation.* The importance of genetic recombination depends on particulate inheritance. It was Weismann who first realized the vast importance of sexual reproduction as a mechanism for the production of almost unlimited genetic recombination (Churchill 1979). Although this factor was rather neglected in the mutationist heyday of Mendelism, it was revived from the 1930s on. Weismann, like the modern evolutionists, was a true follower of Darwin, for whom also the individual was the target of selection. Natural selection would he helpless if there were not an inexhaustible supply of genetic variation in the form of uniquely different individuals.

(5) *Constraints on natural selection.* As discussed above, Weis-

mann strongly emphasized that there are severe developmental and other constraints on the power of selection.

(6) *Mosaic evolution.* Weismann emphasized again and again that not only different components of the phenotype but also different stages in the life cycle vary in their rate of evolution. For instance, in butterflies the caterpillars often evolve faster and along entirely different lines from the imagos (1882:432). As a result, a classification based on larval characters is not at all the same as one based on imagos. Weismann lists instances of mosaic evolution for many groups of insects (1882:481–501). In most cases it results from special adaptations of the larval stages. Indeed, Weismann offers a long list of evolutionary "incongruences" discovered when he compared either larval and adult stages or the corresponding stages of more closely or more distantly related higher taxa (1882:502–519).

(7) *Cohesion of the genotype.* Weismann stated repeatedly that, mosaic evolution notwithstanding, there are limits to the independence of different components of the genotype. Organisms must evolve more or less harmoniously, and a selection pressure on one organ very often results in a selection pressure on some other structure. Or a change in behavior results in the necessity for modification of a structure. In other words, it is the genotype as a whole that responds to the forces of natural selection, a consideration often ignored during the reductionist period of mathematical genetics.

More broadly, in an age of inductionism it was very important that there was someone who had the courage to speculate. Weismann pointed out significant problems and unanswered questions, even in cases where he himself was unable to find the right solution.

Probably no one in the latter part of the nineteenth century comprehended the basic thesis of Darwinism better than Weismann. He was the only one who understood the overwhelming role of natural selection. And he realized that the source of genetic variation was the great unknown in the process of selection and that a detailed theory of inheritance was the need of the hour.

Even though he himself failed to meet that need, the intellectual preparation he gave to the area enabled the Mendelian theory to prosper. Even more than in his own lifetime, Weismann is today considered one of the very few truly outstanding evolutionary biologists.

CHAPTER NINE

Geneticists and Naturalists
Reach a Consensus:
The Second Darwinian Revolution

ADVANCE IN SCIENCE is rarely steady and regular. Nor can it necessarily be described in Thomas Kuhn's terms as a series of revolutions separated by long periods of steadily progressing normal science. Rather, when we study particular scientific disciplines we observe great irregularities: theories become fashionable, others fall into eclipse; some fields enjoy considerable consensus among their active workers, other fields are split into several camps of specialists furiously feuding with one another. This latter description applies well to evolutionary biology between 1859 and about 1940.

The opposition to natural selection continued unabated for some eighty years after the publication of the *Origin*. Except for a few naturalists, there was hardly a single biologist, and certainly not a single experimental biologist, who adopted natural selection as the exclusive cause of adaptation. One would have thought that the rediscovery of Mendel's laws in 1900 would have brought about an immediate change in the attitude toward natural selection, but this was not the case. The findings of genetics made it now quite clear that the genetic material was particulate (hence inheritance could not be blending), and likewise that inheritance was hard (that is, not permitting any inheritance of acquired characters). Yet the leading Mendelians—Bateson, de Vries, and Johannsen—did not adopt natural selection. They ascribed evolutionary change instead to mutation pressure.

By the 1920s most students of evolution belonged to one of three biological disciplines: genetics, systematics, or paleontology. They differed in their interests and in the kind of knowledge they had. The naturalists (both the systematists and the paleontologists) were not sufficiently familiar with the advances in genetics made after 1910, and thus they were arguing against the erroneous evolutionary concepts of the Mendelians as though this were still the viewpoint of genetics. The geneticists, in turn, ignored the rich literature on geographic variation and speciation; consequently, nothing in the evolutionary writings of T. H. Morgan, H. J. Muller, R. A. Fisher, or J. B. S. Haldane could explain the multiplication of species, the origin of higher taxa, or the origin of evolutionary novelties. Furthermore, the two groups dealt with different hierarchical levels: the geneticists with intrapopulational variation at the gene level, the naturalists with the geographic variation of populations and with species. When geneticists and paleontologists, or geneticists and taxonomists, had joint meetings in that period, their respective backgrounds were so different that they were seemingly unable to communicate with one another.

Yet during the 1920s the foundation was laid for an eventual consensus. Not only were blending as well as soft inheritance definitely refuted by the geneticists during this period, but the occurrence of spontaneous mutations was firmly established, something the deterministic physicalists of the nineteenth century had been unable to accept. Also, Morgan and his school as well as Edward East and Erwin Baur found that most mutations had only very small effects on the phenotype and were not at all like the large mutations envisaged by the early Mendelians. In this period the difference between genotype and phenotype was clarified, and it was understood that what is selected are whole genotypes, not individual genes. Therefore, genetic recombination rather than mutation was seen as the immediate source of the genetic variation available for selection. The way selection acts in a population was understood much better when it was connected with these improvements of understanding.

Rather unexpectedly, within a few years, a wide-reaching consensus among geneticists, systematists, and paleontologists was reached. The term "evolutionary synthesis" was introduced by Julian Huxley (1942) to designate the acceptance of a unified evolutionary theory by the previously feuding camps of evolutionists. How could this unexpected establishment of seeming agreement have come to pass so suddenly? Two workshops were organized in 1974 in which an answer to this question was sought (Mayr and Provine 1980). It became clear that the synthesis of the opposing viewpoints was made possible when a number of taxonomists—Sergei Chetverikov, Theodosius Dobzhansky, E. B. Ford, Bernhard Rensch, and I—became acquainted with post-Mendelian genetics (that is, population genetics) and developed an up-to-date Darwinism that combined the best elements of both genetics and systematics. Furthermore, George Gaylord Simpson was responsible for bringing paleontology and macro-evolution into the synthesis. He was able to show that the phenomena studied by the paleontologist—that is, macroevolution, or evolution above the species level—is in every respect consistent with the findings of modern genetics and with the basic concepts of Darwinism. The same was shown independently in Germany by Rensch and in America, for plants, by L. G. Stebbins.

The new consensus was heralded and promoted by Dobzhansky's *Genetics and the Origin of Species* (1937), followed by J. Huxley's *Evolution: The Modern Synthesis* (1942), Mayr's *Systematics and the Origin of Species* (1942), Simpson's *Tempo and Mode in Evolution* (1944), Rensch's *Neuere Probleme der Abstammungslehre* (1947), and Stebbins's *Variation and Evolution in Plants* (1950). These authors have often been referred to as the architects of the evolutionary synthesis. This is a somewhat arbitrary selection; one could have included among the architects also some evolutionists who published prior to 1937, such as Chetverikov, Sumner, Stresemann, Fisher, Haldane, and Wright.

The Nature of the Evolutionary Synthesis

What occurred during the period from 1936 to 1950, when the synthesis took place, was not a scientific revolution; rather it was a unification of a previously badly split field. The evolutionary synthesis is important because it has taught us how such a unification may take place: not so much by any revolutionary new concepts as by a process of house cleaning, by the final rejection of various erroneous theories and beliefs that had been responsible for the previous dissension. Among the constructive achievements of the synthesis was the finding of a common language among the participating fields and a clarification of many aspects of evolution and its underlying concepts.

The period of the synthesis was not one of great innovations but rather of mutual education. Naturalists who had not known it before learned from the geneticists that inheritance is always hard, never soft. There can be no heritable influence of the environment, no inheritance of acquired characters. Weismann's thesis was finally adopted universally more than fifty years after it had first been proposed. Another finding of genetics, its particulate character, was also finally universally adopted. Many naturalists up to that time had divided characters into Mendelian (particulate) ones, which they considered evolutionarily unimportant, and gradual or blending ones, which, following Darwin, they considered to be the true material of evolution.

Acceptance of these two findings of genetics helped in the refutation of the three major evolutionary theories that had competed with natural selection since the publication of the *Origin* (Bowler 1983). These theories, as we have seen, were (1) neo-Lamarckism (the inheritance of acquired characters) and other forms of soft inheritance; (2) autogenetic theories based on the belief in a built-in drive toward evolutionary progress (orthogenesis, nomogenesis, aristogenesis, omega principle); (3) and saltational theories of evolution, which postulated the sudden emergence of drastically new life forms (de Vriesian mutations). Perhaps no author contributed more to the refutation of these three theories than G. G. Simpson, whose *Tempo and Mode* (1944) and *Meaning of Evolution* (1949) consist in large part of evidence disproving them.

But the synthesis was not merely the general acceptance by the naturalists of the principles of theoretical population genetics. To understand the achievements of the synthesis, one must appreciate how typological and saltationist the evolutionary views of the original Mendelians were (Provine 1971). It is a Whiggish misrepresentation of history to equate Mendelism with genetics. To be sure, the two agreed in their rejection of soft and blending inheritance. On the other hand, the evolutionary views of Bateson, de Vries, and Johannsen were rejected during the synthesis (Mayr and Provine 1980). Indeed, neo-Lamarckism seemed to explain evolution better than did the saltationist theories of the Mendelians.

A major achievement of the synthesis, then, was to develop a unified view on the nature of genetic change. Darwin, accepting universal opinion on this subject, had thought there were two kinds of variation: drastic ones, often referred to as sports, and small ones, represented by gradual or quantitative variation. For Darwin, it was gradual variation that was important in evolution. By contrast, the Mendelians insisted that new species originated through drastic mutations. Genetic research by Nilsson-Ehle, East, Castle, and Morgan during the first third of the twentieth century showed clearly that drastic and minutely differing variants were only extremes of a continuously varying spectrum, and that the same genetic mechanism was involved in mutations of all degrees of difference.

This finding had a number of important consequences. It permitted a reconciliation between the Mendelians and those who studied quantitative inheritance. It also permitted the building of a bridge between micro- and macroevolution. Most important, it refuted the credo of essentialism. There is no uniform species essence, but rather each individual has a highly heterogeneous genotype (which varies from individual to individual). The belief in two kinds of variation was still widespread until the 1930s, but the new way of interpreting genetic variation became completely victorious during the synthesis (Sapp 1987).

The realization that gradual variation could be explained in terms of Mendelian (particulate) inheritance also led to the end of any belief in so-called blending inheritance. This was helped by the clear recognition of a difference between genotype and phenotype, as was shown by Nilsson-Ehle, East, and others; a complete "blending" of characters of the phenotype was possible in spite of the discrete particulateness of the underlying genetic factors.

Contributions by the Naturalists

Because of these advances in genetics, it is therefore sometimes asserted that the evolutionary synthesis was merely the application of Mendelian inheritance to evolutionary biology. This formulation ignores two important factors. One is that the geneticists, particularly the experimental and the mathematical ones, had to acquire just as much from the naturalists as the naturalists had to acquire from the geneticists. And second, the conceptual framework of evolutionary biology was greatly enriched by concepts and facts from natural history that were conspicuously absent in the writings of the geneticists.

For example, evolution, defined by the geneticists as "a change of gene frequencies in populations," was visualized essentially as a strictly temporal phenomenon. This is reflected in Muller's statement: "Speciation represents no absolute stage in evolution, but is gradually arrived at, and intergrades imperceptibly into racial differentiation beneath it and generic differentiation above" (Muller 1940:258). Throughout their writings the geneticists concentrated almost completely on the temporal component in evolution. The paleontologists, by necessity thinking in terms of vertical sequences, likewise confined themselves to the study of vertical evolution until they merged their thinking with the horizontal tradition of the naturalists (Eldredge and Gould 1972). One of the most important contributions of the naturalists, the heirs of Darwin, was to bring geographical thinking into the synthesis. The problem of the multiplication of species, the existence of polytypic species, the biological species concept, the role of species and speciation in macroevolution, and many other evolutionary problems can be dealt with only by invoking geographical evolution.

The incorporation of the geographical dimension was of particular importance for the explanation of macroevolution. Paleontologists had long been aware of a seeming contradiction between Darwin's postulate of gradualism, confirmed by the work of population genetics, and the actual findings of paleontology. Following phyletic lines through time seemed to reveal only minimal gradual changes but no clear evidence for any change of a species into a different genus or for the gradual origin of an evolutionary novelty. Anything truly novel always seemed to appear quite abruptly in the fossil record. During the synthesis it became clear that since new evolutionary departures seem to take place almost invariably in localized isolated populations, it is not surprising that the fossil record does not reflect these sequences. A purely vertical approach is unable to resolve the seeming contradiction.

An equally important contribution made by the naturalists was the introduction of population thinking into genetics. Mendelism was strongly typological—the mutation versus the wild type. And even later, in the thinking of many geneticists, not only Morgan and Goldschmidt but even Muller in his search for the perfect genotype, and R. A. Fisher, one can detect a strong essentialistic component. Population thinking, with its emphasis on the uniqueness of every individual in the population, was brought into genetics by Chetverikov and his students (including Timofeeff-Ressovsky), by Dobzhansky, and by Baur. With few exceptions (for example, polyploidy), every evolutionary phenomenon is simultaneously a genetic phenomenon and a populational one.

Finally, the naturalists, or at least some of them, attempted to replace the strictly reductionist formulation of most geneticists by a more holistic approach. Evolution, they said, is not merely a change in the frequency of genes in populations, as the reductionists asserted, but is at the same time a process relating to or-

gans, behaviors, and the interactions of individuals and populations. In this holistic attitude the naturalists agreed with the developmental biologists.

The Triumph of Natural Selection

The synthesis was a reaffirmation of the Darwinian formulation
that all adaptive evolutionary change is due to the directing force
of natural selection on abundantly available variation. Today we
are so completely used to the Darwinian formulation that we are
apt to forget how different it is from the evolutionary explanations of Darwin's opponents.

The naturalists, in the tradition of Darwin, had been the
staunchest defenders of natural selection from the very beginning, but like Darwin almost all of them tended to believe simultaneously in a certain amount of soft inheritance. Now that this
form of inheritance was decisively refuted, the strongest supporters of natural selection were found among the naturalists.

There is, as yet, no good history of the acceptance of natural
selection by the geneticists. Chetverikov, Timofeeff-Ressovsky,
and Dobzhansky got it from a strong Russian tradition (Adams
1980). In no other country did natural selection have as widespread a base of support as in Russia. But for ideological and political reasons this highly successful genetic establishment was
wiped out by Stalin's regime and in its place was installed the
charlatan Lysenko and his henchmen. In England there was a
strong selectionist tradition at Oxford (Lankester, Poulton, and
so on), but the classical Mendelians had no use for selection, and
in the United States Morgan's thinking was very much in that
tradition. A second tradition in the United States, however—
centered in Harvard's Bussey Institution (Castle, East, Wright)—
apparently adopted selection without reservation. In France,
Lwoff at the Institut Pasteur seems to have been the first consistent selectionist, followed by Ephrussi, L'Héritier, and Teissier.
But even now strict selectionists seem to be in the minority

among French biologists. In every country except the USSR selectionism gradually grew in strength in the years after the synthesis.

It is understandable that in the early stages of the synthesis the universal presence of natural selection should have been emphasized strongly, since a considerable number of Lamarckians still existed among the older evolutionists. However, as soon as this stage passed, there developed a trend toward recognition of other factors. Where the modern biologist perhaps differs from Darwin most is in assigning a far greater role to stochastic processes than did Darwin or the early neo-Darwinians. Chance plays a role not only during the first step of natural selection—the production of new, genetically unique individuals through recombination and mutation—but also during the probabilistic process that determines the reproductive success of these individuals. All sorts of constraints forever prevent the achievement of "perfection." Even though natural selection is indeed an optimization process, the existence of numerous opposing influences makes optimality quite unachievable.

The unification of evolutionary biology achieved by the synthesis painted its picture in bold strokes: Gradual evolution is due to the ordering of genetic variation by natural selection, and all evolutionary phenomena can be explained in terms of the known genetic mechanisms. This was an extreme simplification, considering that processes in organismic biology are usually highly complex, often involving several hierarchical levels and pluralistic solutions. The task of evolutionary biology after the synthesis of the 1940s was to convert the coarse-grained theory of evolution into a fine-grained, more realistic one. No longer constrained to a defense of Darwinism, the followers of the evolutionary synthesis, in the course of the more detailed analysis, began to tackle differences that still existed, not only between the reductionist tendencies of the geneticists and the organismic viewpoint of systematists and paleontologists but also concerning other aspects of evolutionary theory.

CHAPTER TEN

New Frontiers in Evolutionary Biology

J UST AS IN THE DECADE after the rediscovery of Mendel's rules, since about 1970 the claim has been made increasingly often that "Darwinism is dead." I shall not deal at all with the attacks by creationists, based on ideological commitments, since their arguments have been decisively refuted by Futuyma (1983), Kitcher (1982), Montagu (1983), Newell (1982), Ruse (1982), Young (1985), and several other authors. Claims that Darwinism is obsolete have been made in numerous articles and books also by several nonbiologists, whose arguments, though nonreligious, are based on such ignorance of evolutionary biology that it is not worthwhile to provide references to their writings. More disturbing are the similar, although somewhat more muted, claims that have been made by some knowledgeable biologists— even evolutionary biologists. These include Gould (1977; 1980), Eldredge (1985), White (1981), and Gutmann and Bonik (1981). Different critics have singled out different aspects of the synthesis as particularly vulnerable. It has been claimed, for example, that:

(1) The findings of molecular biology are incompatible with Darwinism.

(2) The new research on speciation shows that other modes of speciation are more widespread and more important than geographic speciation, which the neo-Darwinians claim is the prevailing mode.

(3) Newly proposed evolutionary theories, like punctuationism, are incompatible with the synthetic theory.

(4) The synthetic theory, owing to its reductionist viewpoint, is unable to explain the role of development in evolution.

(5) Even if one rejects the reductionist claim of the gene as the target of selection, Darwinism, by considering the individual the target of selection, is unable to explain phenomena at hierarchical levels above the individual, that is, it is unable to explain macroevolution.

(6) By adopting the "adaptationist program," and by neglecting stochastic processes and constraints on selection, particularly those posed by development, the evolutionary synthesis paints a misleading picture of evolutionary change.

These highly diverse criticisms range from the extreme view that Darwinism as a whole has been refuted to the milder opinion that the synthesis is too narrowly adaptationist or that the concept of speciation has to be thoroughly revised. One by one these various criticisms have been refuted by Ayala (1981; 1983), Stebbins and Ayala (1981), Grant (1983), Maynard Smith (1982; 1983), Mayr (1984), A. Huxley (1982), Levinton (1988), and others.

The aspect of almost all of these objections that I wish to address here is an evident misunderstanding of the theory that emerged from the evolutionary synthesis. Often this misunderstanding arises from the assumption that the most extreme reductionist version of the synthesis, as represented by some of the mathematical population geneticists, is the basic dogma of the synthesis. When critics propose that the conclusions of the synthesis be replaced by a more modern view of evolution, one finds practically without exception that these supposedly novel views have been maintained all along by Rensch, myself, and several other representatives of the synthesis. Almost all the critics of the synthesis, I am sorry to say, have quite conspicuously misrepresented the views of its leading spokesmen.

To take one example, the critics continually criticize the claim that "Darwinian evolution is due to the natural selection of random mutations." This criticism completely ignores the fact that from Darwin on to the 1980s the individual as a whole was considered the target of selection for the organismic biologists, and therefore recombination and the structure of the genotype as a whole were viewed as being far more important for evolution than mutational events at individual loci. Furthermore, the critics completely misinterpret the word "random." The term, when applied to variation, means that it is not in a response to the needs of the organism.

Opponents of the synthesis consistently confound three schools of Darwinism: (1) neo-Darwinism, a term coined by Romanes in 1896 to designate "Darwinism without an inheritance of acquired characters"; (2) early population genetics, a strongly reductionist school that defined evolution as the modification of gene frequencies by natural selection; and (3) the holistic branch of the synthesis, which continued the traditions of Darwin and the naturalists while accepting the findings of genetics.

The thinking of the reductionists was strongly influenced by R. A. Fisher, and this school has therefore sometimes been designated as Fisherian Darwinism. It is clearly the primary target of the opponents of the synthesis, but these critics confuse matters when they designate the reductionist school as neo-Darwinism or imply that it includes people like Huxley, Dobzhansky, Wright, Rensch, or myself, all of whom distinctly rejected the reductionist conclusions of the Fisherian school.

Darwinism is not a simple theory that is either true or false but is rather a highly complex research program that is being continuously modified and improved. This was true before the synthesis, and it continues to be true after the synthesis. Table 2 lists many of the significant stages in the modification of Darwinism that one might recognize. Yet recognizing such seemingly discontinuous periods is in many respects an artificial enterprise. To take but a few examples, the prevalence of particulate inheritance was obvious after 1886 but was not adopted until after 1900; the emphasis on the role of diversity in evolution was stressed by naturalists from Darwin on but was almost entirely ignored by the Fisherians; the naturalists, for their part, rejected the beanbag genetics of the reductionists and during the post-synthesis period continued their holistic tradition of emphasizing the individual as the target of selection. In short, each of these periods was heterogeneous to some extent, owing to the diversity in the thinking of different evolutionists. Most critics who have attempted to refute the evolutionary synthesis have failed to recognize this diversity of views and thus have succeeded in refuting only the reductionist fringe of the Darwinism camp. Their failure to appreciate the complexity of the evolutionary synthesis has led them to paint a picture of that period that is at best a caricature.

Another error made by most opponents of the synthesis is a failure to differentiate between proximate and evolutionary causations. For Darwin and all of his holistic followers, selection starts at fertilization and continues through all embryonic and lar-

TABLE 2
Significant stages in the modification of Darwinism.

Date	Stage	Modification
1883; 1886	Weismann's neo-Darwinism	End of soft inheritance; diploidy and genetic recombination recognized
1900	Mendelism	Genetic constancy accepted and blending inheritance rejected
1918–1933	Fisherism	Evolution considered to be a matter of gene frequencies and the force of even small selection pressures
1936–1947	Evolutionary synthesis	Population thinking emphasized; interest in the evolution of diversity, geographic speciation, variable evolutionary rates
1947–1970	Post-synthesis	Individual increasingly seen as target of selection; a more holistic approach; increased recognition of chance and constraints
1954–1972	Punctuated equilibria	Importance of speciational evolution
1969–1980	Rediscovery of sexual selection	Importance of reproductive success for selection

val stages. A Darwinian is truly puzzled when he reads in a critique by an embryologist that "development . . . comes to the fore as a problem unintelligible within neo-Darwinism." What aspect of development is this author talking about? If he is speaking of the translation of the genetic program into molecular chains of events during ontogeny, he is talking about proximate causations. Their study, indeed, has never been the job of the evolutionary biologist. But many other aspects of development do raise questions concerning evolutionary causations and these have been of interest to evolutionists from Darwin on. They are of concern to the evolutionist, first, because each stage of development is a target of selection, and particularly so when the developmental stages (larvae) are free-living. Second, because embryonic stages may themselves serve as "somatic programs" in development (see below), such stages tend to become highly conserved in evolution (for instance, the gill arch stage of the tetrapod embryo). Such highly conserved stages are often helpful in the reconstruction of the phylogeny (recapitulation). No Darwinian will ever question the importance of development in evolution, but evolutionary interpretation is constrained by the extent to which the proximate causations of development have been elucidated by the embryologists. To undertake a study of such proximate causations is not the task of the evolutionist.

Curiously, the objection is sometimes raised that the evolutionary synthesis can shed light neither on the level of the gene nor on trans-specific levels because it is concerned with individuals (as targets of selection) and with populations (as incipient species). What is needed, it is claimed, is a hierarchical approach that is not found in either neo-Darwinism or the evolutionary synthesis. That this objection is without basis has been shown by Grant (1983:153), Stebbins and Ayala (1981), and other defenders of the synthesis. Even Simpson's interpretation of evolution was strongly hierarchical (Laporte 1983). No one has shown as clearly as Mayr (1963:621) that the species is the unit of action in macroevolution.

And finally, for White (1981), the synthesis was premature,

"because there was no molecular biology at the time and both the chemical nature of the hereditary material and the architecture of eukaryote chromosomes were yet unknown." This is like saying that the first Darwinian revolution was premature because genetics had not yet been founded. Any scientific revolution or synthesis has to accept all sorts of black boxes, for if one had to wait until all black boxes are opened, one would never have any conceptual advances.

The Evolutionary Synthesis as Unfinished Business

The healthy turmoil that currently characterizes evolutionary biology should not be viewed as a death struggle but rather as the sort of lively activity one will find in any healthy and advancing branch of science. Yet it must be admitted that, even though the refutation of the major anti-Darwinian theories during the synthesis drastically narrowed down the variation of evolutionary theory, some well-defined differences among the Darwinians still existed into the post-synthesis period, and some of these differences are still with us fifty years later.

For the geneticists—or at least all those influenced by Muller, Fisher, and Haldane—the gene continued to be the primary target of selection, and most genes were believed to have constant fitness values. The whole problem of the origin of organic diversity (that is, the multiplication of species) was minimized by this school, if not ignored. Most of those who called themselves population geneticists worked with single, closed gene pools. Even Wright did not come to grips with the problem of the multiplication of species in his shifting-balance theory, nor with the macroevolutionary problems generated by speciation.

By contrast, those who had come to evolutionary biology from systematics or some other area of natural history considered evolution to be a populational problem, and the whole (potentially reproducing) individual to be the target of selection. For them, the multiplication of species was the pathway to the solution of problems of macroevolution. Despite their tendency to think in terms of phenotypes, they eventually came to view the genotype as a system of gene interaction—that is, they recognized the cohesion of the genotype—and tended from the very beginning to deal with evolution hierarchically. For Simpson, Rensch, Huxley, and Mayr, evolution was not a change in gene frequencies but the twin processes of adaptive change and the origin of diversity.

By no means are all current intra-Darwinian controversies remnants of the old geneticist-versus-naturalist feud. There are also differences among the geneticists concerning the relative frequency of neutral mutations and the amount of variation due to balance. Among the paleontologists there are disputes as to whether or not phyletic gradualism can give rise to higher taxa. And among other evolutionists there are disagreements on the validity of group selection, the extent of adaptation, the role of competition, the frequency of sympatric speciation, how continuous or punctuated evolution is, to what extent all components of the phenotype reflect ad hoc selection, what proportion of speciation is allopatric, what the target of selection is, how much genetic variation is stored in populations, and to what extent the new findings of molecular biology require a revision of current theory. However, regardless of ultimate outcome, none of these disagreements affects any of the basic principles of Darwinism.

Although rarely completely wrong, the conclusions supported by the followers of the evolutionary synthesis were often incomplete and rather simplistic. Two kinds of processes, in particular, were often inadequately considered: (1) multiple simultaneous causations, and (2) pluralistic solutions. Let me give some examples of simultaneous causations. In all selection phenomena—and selection is of course an antichance process—chance phenomena also occur simultaneously. Or, to give another example, speciation is never merely a matter of genes or chromosomes but also of the nature and geography of the populations in which the genetic changes occur. Geography and the genetic changes of populations affect the speciation process simultaneously.

Far more serious was a frequent neglect of the pluralism of evolutionary phenomena. Darwin was well aware of the pluralism of evolutionary processes. The more one studies these processes, the more one is impressed by their diversity. Many controversies in evolutionary biology arise because of the inability of certain authors to appreciate this diversity. There seem to be multiple solutions to almost any evolutionary challenge. During speciation, pre-mating isolating mechanisms originate first in some groups of organisms; in others, post-mating mechanisms originate first. Sometimes geographic races are phenotypically as distinct as good species, yet not at all reproductively isolated. On the other hand, phenotypically indistinguishable species (sibling species) may be fully isolated reproductively. Some species are extraordinarily young, having originated only 2,000 to 10,000 years ago, while others have not changed visibly in 10 to 50 million years. Polyploidy or asexual reproduction is important in some groups and totally absent in others. Chromosomal restructuring seems to be an important component of speciation in some groups, such as the morabine grasshoppers of Australia, but seems not to occur in others. Interspecific hybridization is frequent in some groups but is rare or absent in others. Some groups speciate profusely; in others speciation seems to be a rare event.

Just because there is little gene flow in certain species, one cannot conclude that gene flow is irrelevant in all species. In fact, the amount of gene flow seems to differ greatly even in closely related species. The absence of major genetic reconstruction in many founder populations does not prove that genetic revolutions can never take place. Parapatry is frequent in groups in which post-zygotic isolating mechanisms evolve first when pre-zygotic ones are still very incomplete; however, this does not justify explaining all cases of parapatry by the same mechanism. One phyletic lineage may evolve very rapidly while others, even closely related ones, may experience complete stasis for many millions of years. In short, there are several possible solutions for many evolutionary challenges, but all of them are compatible with the Darwinian paradigm.

Most evolutionists, particularly those who work on a single group of organisms, tend to neglect the extraordinary pluralism of evolution. As François Jacob (1977) has so rightly said, evolution is a tinkerer, and in a given situation makes use of that which is most readily available. One can take almost any evolutionary phenomenon and show how greatly it differs among the different groups of animals and plants. The lesson one learns from this is that sweeping claims are rarely correct in evolutionary biology. Even when something occurs "usually," this does not mean that it must occur always. One must remember the forever-present pluralism of evolutionary processes.

Almost every scientific theory is continuously in need of revision and supplementation, yet these changes do not necessarily touch the true core of the theory. The Mendelian theory of inheritance is an apt illustration of this. When it was rediscovered in 1900, it was expressed in the form of three laws. Within just a few years two of these laws, those of dominance and independent assortment, were found not to be universally valid. Still, no one claimed that Mendel's theory had therefore been refuted. The continuing minor revisions and supplementations of the Darwinian theory, including the version given to it during the synthesis, do not qualify as refutations either.

The architects of the evolutionary synthesis have been accused by some critics of claiming that they had solved all the remaining problems of evolution. This accusation is quite absurd; I do not know of a single evolutionist who would make such a claim. All that was claimed by the supporters of the synthesis was that they had arrived at an elaboration of the Darwinian paradigm which seemed to be sufficiently robust not to be endangered by the remaining puzzles. No one denied that there were many open questions, but there was the feeling that no matter what answer would emerge in response to these questions, it would be consistent with the Darwinian paradigm. Up to now, it seems to me, this confidence has not been disappointed.

The Impact of Molecular Biology on Evolutionary Theory

Early in the history of molecular biology there was a widespread feeling that its new discoveries might necessitate a complete rewriting of evolutionary theory. Considering how quickly molecular biology developed into a large and powerful field of its own, a call for an integration of this new field with classical evolutionary biology was to be expected. As White (1981) said, "We once again urgently need a new synthesis of two traditions—those of evolutionary and molecular biology." Other molecular biologists went even further and stated that the findings of molecular biology had already refuted much of accepted Darwinism. Thus, it was claimed, "Many of the observations [of molecular biology] (inducible mutation systems, rapid genomic changes involving mobile genetic elements, programmed changes in chromosome number) challenge the most fundamental assumptions which evolutionary theories make about the mechanisms of hereditary variations and the fixation of genetic differences" (Shapiro 1983).

Discoveries that seem to be in conflict with the picture of classical genetics are made daily in molecular biology. Perhaps none was as startling as the discovery that genes are highly complex systems consisting of exons, introns, and flanking sequences, and that there are numerous kinds of genes; some have seemingly no function at all, while several functional classes can be distinguished among the active genes. But have any of these discoveries required a revision of Darwinism? I do not think so.

There is no question that molecular biology has given us numerous great new insights into the working of evolutionary causes, particularly the production of genetic variation. Gratifyingly, in most cases they were a confirmation or elaboration of existing views. Let me merely mention some of the most important molecular discoveries.

(1) The genetic program does not by itself supply the building material of new organisms, but only the blueprint for making the phenotype.

(2) The pathway from nucleic acids to proteins is a one-way street. Proteins and information that they may have acquired are not translated back into nucleic acids.

(3) Not only the genetic code but in fact most basic molecular mechanisms are the same in all organisms, from the most primitive prokaryotes up.

(4) Many mutations (changes in the base pairs) seem to be neutral or near-neutral, that is, without noticeable effect on the selective value of the genotype, but this varies from gene to gene (see below).

(5) A critical comparative analysis of molecular changes during evolution provides a very large number of pieces of information suitable for the reconstruction of the phylogeny. This is particularly useful if the morphological evidence is indecisive. However, molecular characters are also vulnerable to homoplasy—parallel or convergent production of the same character or phenotype.

Interestingly, up to the present time, none of these findings has required an essential revision of the Darwinian paradigm. Moreover, the relation between molecular and evolutionary biology has not been entirely one-sided. Darwinian thinking made a major contribution to the evolution of biochemistry into molecular biology. The study of the phylogeny of molecules and the search for the selective significance of molecular structure have greatly enriched molecular biology. No longer does one encounter in the literature of molecular biology any thorough study of a particular molecule or group of molecules that is not at the same time concerned with the evolutionary explanation of the molecular structures encountered during that analysis.

NEUTRAL EVOLUTION

In classical population genetics it was assumed that all mutations were either somewhat favorable and would then be incorporated into the genotype (unless lost by random events), or more or less deleterious, and then eliminated whenever becoming homozygous. This assumption is, for instance, reflected in the writings of R. A. Fisher and H. J. Muller. However, genetic variation found in nature actually seemed to be larger than could be accounted for by such a process of relatively rapid elimination of variation. This induced Dobzhansky to hypothesize that heterozygotes were sometimes superior ("overdominant") to the homozygotes and that through such "balanced selection" the holding capacity of the gene pool was considerably increased.

In 1966 when enzyme genes were studied in natural populations by the new method of electrophoresis (Hubby and Lewontin 1966; Harris 1966), it was found that variant genes (alleles) were enormously frequent. For this and other reasons Kimura (1968) and King and Jukes (1969) proposed a theory of "neutral evolution" in which it was claimed that many, perhaps most, amino acid and nucleotide substitutions in evolution are random fixations of neutral or nearly neutral mutations. A new allele (or base-pair substitution) is called neutral if it is functionally equivalent to the allele (or base pair) it replaces and does not change the fitness of an organism. Although, this theory was at first vigorously opposed by most evolutionists, including myself, the high frequency of "neutral" base-pair replacements is now well established. On the other hand, the selective significance of numerous alleles that had been considered neutral by neutrality enthusiasts has also been established (for instance by Nevo 1983).

One major reason for the heat of the argument between neutralists and their opponents was that they had a different interpretation of the target of selection. The neutralists are reductionists, and for them the gene—more precisely the base pair—is the target of selection. Hence, any fixation of a "neutral" base pair is a case of neutral evolution. For the Darwinian evolutionists, the individual as a whole is the target of selection, and evolution takes place only if the properties of the individual change. A replacement of neutral genes is considered merely evolutionary noise and irrelevant for phenotypic evolution. Whichever interpretation one adopts, molecular researches have revealed that the number of neutral changes seem to be far greater than the number of gene changes that have adaptational significance.

The more important a gene or part of a gene is functionally, the less likely it is that a mutation could occur that would improve it. Nearly all mutations of such a gene would be deleterious and would be selected against. One potential significance of neutral evolution about which we are still ignorant is the possibility that some new alleles produced by neutral mutation may later in evolution have a positive selective value on a different genotypic background.

The Darwinian wonders to what extent it is legitimate to designate as evolution the changes in gene frequencies caused by nonselected random fixation. In some of the older (particularly nineteenth-century) literature on evolution, one finds discussions on how to discriminate between evolution and mere change. There it was pointed out that the continuing changes in weather and climate, the sequence of seasons of the year, the geomorphological changes of an eroding mountain range or a shifting river bed, and similar changes do not qualify as evolution. Interestingly, the changes in nonselected base pairs and genes are more like these nonevolutionary changes than they are like evolution. Perhaps one should not refer to non-Darwinian evolution but rather to non-Darwinian changes during evolution.

Punctuated Equilibria

Paleontologists, from Darwin's days on, have pointed out that continuous series of fossils either remain unchanged in time or show only minor changes in size or proportions. All major evolutionary changes seem to occur rather abruptly, not connected by intermediates with the preceding fossil series. It was this observation that induced Schindewolf and several other paleontologists to propose saltational evolution. How can an adherent of Darwinian gradual evolution explain the riddle of these gaps? The theory of peripatric speciation, that is speciation in geographically isolated populations, permitted Mayr (1954, 1963) to suggest that such inbreeding incipient species are sometimes the scene of particularly rapid genetic turnover, which would leave no evidence in the fossil record, owing to the geographical restrictedness and short duration of the founder populations.

On this foundation Eldredge and Gould (1972) erected the theory of punctuated equilibria, stating that most major evolutionary events take place during short bouts of speciation and that successful new species, after they become widespread and populous, enter a period of stasis, lasting sometimes for many millions of years, during which they show only minimal change. Such speciational evolution, because it occurs in populations, is gradual in spite of its rapid rate and therefore is in no conflict whatsoever with the Darwinian paradigm. How often such major changes in founder populations occur and what percentage of new species enter a subsequent period of stasis are still controversial, however.

Sociobiology

Perhaps the most important development in evolutionary biology is the adoption of a largely evolutionary approach in almost all branches of biology. For instance, the ecologist asks to what extent an ecosystem is the result of such evolutionary forces as competition and predation? By what kind of selection pressures is the partitioning of resources controlled? Indeed, every ecological problem is subjected to a reconstruction of the selection factors seemingly involved, and to a study of their possible adaptive significance. The same is true for the field of animal behavior. Not only are the physiological causations of behavior studied but also the adaptive significance of each kind of behavior, and the selective agents that control behavior in relation to other adaptations of a species.

It is the merit of Edward O. Wilson's magnificent work *Sociobiology: The New Synthesis* (1975) to have pointed out that an evolutionary approach is particularly important in the study of social behavior. Wilson defined sociobiology "as the systematic study of the biological basis of all social behavior." Unfortunately, the slight ambiguity of the word "biological" resulted almost immediately in a heated controversy. For Wilson, "biological" meant that a genetic disposition makes a *contribution* to social behavior. For his politically motivated opponents, "biological" meant that social behavior is "genetically determined." Of course, we humans would be mere genetic automata if all of our actions were strictly determined by genes. Everyone agrees that this is not the case. And yet we also know, particularly from studies of twins separated at birth and adopted into different families, and from other adoption studies, that a remarkably large proportion of our attitudes, qualities, and propensities is affected to some degree by our genetic heritage. The modern biologist is far too sophisticated to want to revive the old polarized nature–nurture controversy; we know that almost all human traits are influenced both by inheritance and by the cultural environment (Alexander 1979). The most important point made by Wilson is that in many respects the same problems are encountered in human-behavior studies as in animal studies. Likewise, many of the answers that seem to explain animal behavior are applicable in the study of human behavior.

Although the term "sociobiology" was coined only in 1975, the study of the phenomena now often bracketed together under this term had begun long before; indeed some of these problems had already been constructively analyzed by Darwin. The number of problems in this field is legion, but I will mention only a few. Perhaps the core problem is the question of the evolution of altruism. Selection by its nature is a thoroughly selfish process, measured only in terms of the reproductive advantage it gives to an individual. The great question, then, is how can true altruism evolve? An altruistic act has been defined by Trivers (1985) as an act "that benefits another organism at a cost to the actor, where cost and benefit are defined in terms of reproductive success." This leads to the question: How could natural selection favor an act that is detrimental to the actor?

One suggestion was that it could develop by way of so-called "reciprocal altruism," by which actor A performs an altruistic act toward individual B in the expectation that this individual will reciprocate by performing in turn an altruistic act toward A. To me, application of the term altruism to such an exchange of favors is misleading, because such behavior is not detrimental to the respective actors if there is appropriate reciprocation.

Two other mechanisms have been suggested as explanations for the evolution of altruism. As far back as 1932 J. B. S. Haldane suggested that altruistic acts toward close relatives might favor the survival and spread of those of the altruist's genes that are shared by these relatives. Since fitness is measured in terms of the survival of the genes of an individual, it is legitimate to refer to the successful transfer of one's genes to one's descendants and other relatives as "inclusive fitness." The process of contributing toward the survival of one's close relatives (who share portions of one's genotype) has been referred to as "kin selection."

Parental care, of course, is a special kind of altruism that leads to an enhancement of reproductive success. There is an extended literature on the interaction of mates with each other, with their offspring, and with other relatives, all of it leading to kin selection. Properly speaking, this is merely part of the enormously important topic of "selection for reproductive success." There are factors that favor kin selection, and others that oppose it. The final outcome is a compromise between these conflicting selection pressures. Maynard Smith's (1982) application of game theory to the subject of altruism and selfish behavior has been particularly illuminating.

Strictly speaking, kin selection operates only in groups of close relatives because the genetic advantage of helping another individual decreases rapidly with genetic distance. However, there is another mode by which altruistic behavior can be favored in evolution. In some groups, consisting of both relatives and unrelated members, certain modes of social facilitation have become established that help the entire group. A typical example is the existence of individuals that serve as "watchdogs" while the rest of the group is feeding. Such a monitor, when giving warning calls to alert the other members of his group to a predator, may expose himself to increased vulnerability, but this altruistic behavior benefits the survival and reproductive success of his group as a whole.

Such social "ethics" exist in a rather rudimentary way in many living organisms, and are described in great detail in Wilson's *Sociobiology*. They are of infinitely greater importance in the human species, where every cultural group has its code of ethics which, in the long run, determines the survival and ultimate success of the group. In the human species, quite clearly, one must distinguish two components of ethical behavior, an ancestral one based on inclusive fitness, particularly parental care, and a cultural one, codified in all civilizations in laws and religious dogmas. Undoubtedly, a genetic propensity for accepting and maintaining such cultural prescriptions is favored by selection (Waddington 1960), but the contents of the ethical repertory are acquired in the individual's lifetime and are not fixed genetically.

The study of groups exhibiting altruistic behavior has revived a long controversy about the validity of so-called group selection. Some authors have postulated that not only individuals but entire groups may sometimes serve as the target of selection. We now understand that this claim is valid only for groups with a fitness value that is greater than the arithmetic mean of the fitness values of the individuals of which it is composed. There are two such groups—those consisting of relatives, where inclusive fitness contributes to the fitness of the group, and groups of nonrelatives which practice social facilitation or various forms of mutual help. The controversy as to the validity of group selection has been resolved by a careful discrimination between these various kinds of groups. Some species are, while others are not, subject to group selection.

Questions at the Frontier

If someone were to ask me which frontiers of evolutionary biology are likely to see the greatest advances in the next ten or twenty years, I would say elucidation of the structure of the genotype and of the role of development. The simplistic reductionist view of the relation among genotype, development, and evolution is that each gene is translated into a corresponding component of the phenotype and that the contributions of these components to the fitness of the resulting organism determine the selective value of the genes. There is undoubtedly some truth in this view, but it greatly oversimplifies the actual connections. This had been realized all the way back to Darwin's day. Darwin himself spoke of the mysterious laws of correlation, and Haeckel converted the Meckel-Serres Law into the evolutionary principle of recapitulation. Again, this principle was a vast oversimplification, but it has a correct nucleus. The fact that the land-living vertebrates go through a gill arch stage in their development is a powerful clue for their descent from aquatic ancestors, to give only one example. Both of these generalizations call attention to holistic interactions. By contrast, the belief of some reductionists that the role of genes is exhaustively described by the largely independent contribution of each gene to some particular aspect of the phenotype neglects the fact that the genotype is a complexly interacting system.

Each year one or two books or symposia volumes are published on the relation between genotype, development, and evolution, and it is quite impossible in this short chapter even to outline the basic problems. To make matters worse, though the questions are increasingly better understood, no satisfying answers have yet been found for most of them. When answers have been suggested, they often contradict one another. Instead of attempting to present a detailed account of the problems and the opposing attempts to solve them, I will satisfy myself by merely indicating some of the general directions of future research.

DOMAINS OF THE GENOTYPE

We know now that there are different classes of genes, and they play not only different roles in ontogeny but also in evolution. Furthermore, certain genes seem to be tied together into functional units and seem to control development as units. They seem to represent well-circumscribed domains that lend a hierarchical structure to the genotype. The existence of such domains is not necessarily in conflict with Mendelian segregation. How the developmental and evolutionary conservation of such domains is effected is not yet understood, though the discovery of the widespread occurrence (from yeasts to mammals) of homeoboxes (Robertson 1985) and the study of complete sets of immune genes reveal possibilities.

At the present time the existence of such domains is primarily indicated by indirect evidence. Among a number of categories of such evidence, I would like to discuss in some detail one particular phenomenon, called mosaic evolution.

For a physicist, one of the most important parameters of any process is its rate. In the endeavor to make evolutionary biology as similar to the physical sciences as possible, evolutionists have attempted from 1859 on to determine evolutionary rates (Simpson 1944, 1953). Alas, this has been a rather frustrating experience. Rates of evolution, speciation, and extinction may differ by several orders of magnitude in different organisms and under different circumstances. Perhaps most disturbing was the discovery that organisms do not evolve as harmonious wholes but that different characters (components of the phenotype) and domains of the genotype may evolve at highly different rates. This was already known to Lamarck and Darwin and has since been confirmed by numerous evolutionists, particularly the paleontologists. *Archaeopteryx* is an apt illustration. In some of its characteristics (feathers and wings) it is already a typical bird, in others (teeth and tail) it is still a reptile, and in still other characters it is more or less intermediate between the two classes. A. C. Wilson (1974) has pointed out that frogs are morphologically highly conservative, and this is also true for their chromosomes, but that different phyletic lines of frogs seem to diverge in their enzyme genes at about the same rate of divergence as mammals, which evolve very rapidly in all three types of characters. What this suggests is that different domains of the genotype seem to evolve independently to a considerable degree. This is well illustrated, for instance, in a comparison of man with his anthropoid relatives. Over all, man is most similar to the chimpanzee, but in a few characters man's genotype is more similar to that of the gorilla or even the orangutan.

The term "mosaic evolution" has been used for the independent evolution of different domains of the genotype. At the molecular level it was found not only that different molecules have different evolutionary rates but also that the same molecule may change its rate at different stages in the evolution of a phyletic line. Whenever there is a period of major adaptive radiation, it is accompanied by a highly accelerated rate of molecular evolution, later followed by a deceleration to slow change. The so-called molecular clock may keep very different times for different molecules and may change its rate within a phyletic line. As Goodman (1982:377) has correctly pointed out, none of these findings on changes in molecular timing invalidates the Darwinian thesis that natural selection is the principal force directing the evolution of molecules and organisms. Mosaic evolution, however, must be carefully considered in the construction of phylogenetic trees.

The opposite of the independent evolution of domains of the genotype is the integration of these domains and the seeming cohesion of the genes that belong to any one of these domains. Just exactly what controls this cohesion is still largely unknown, but its existence is abundantly documented. The almost total integrity of the gene complex that controls the number of extremities in tetrapods, insects, and arachnids is one of literally hundreds of examples. During the pre-Cambrian period, when the cohesion of the eukaryote genotype was still very loose, seventy or more distinct morphological types (phyla) formed. Throughout evolution there has been a tendency for a progressive "congealing" of the genotype, so that deviation from a long-established morphological type has become more and more difficult. This is one of the well-known constraints in evolutionary change, and one of the reasons why natural selection has such limited leeway. When we speak of the congealing of the genotype or, if we prefer, of the progressive integration of the genotype, we are simply using words to cover up our ignorance. Nevertheless, the evidence indicates that the older a taxon is, the more difficult it is to escape from the straitjacket of a highly integrated (congealed) genotype. This is why not a single new morphological type (phylum) has originated since the Cambrian, over 500 million years ago.

At first sight this conclusion would seem to be contradicted by the adaptive radiations that have occurred in so many phyletic lines. After a stasis of 100 million years or longer a higher taxon may suddenly experience a burst into numerous new taxa, as happened to the mammals in the Paleocene and Eocene. However, most of these were ad hoc adaptations and did not really deviate very far from the ancestral *Bauplan* or structural type. So far as I can judge, the same is true for most of the other well-known radiations. The songbirds, for instance, have produced over 5,000 living species since their origin, but except for their plumage and bills, they exhibit hardly any deviation from the standard *Bauplan*.

The power of these constraints is evident at every hierarchical level, sometimes even at the species level, as documented by the frequency of sibling species (which are reproductively isolated but morphologically identical). Indeed, most geographic variation deals merely with minor quantitative changes in characters, without affecting the genotype appreciably. Any restructuring, when it occurs, apparently takes place almost always in peripherally isolated founder populations.

Evidence for increasing or decreasing cohesion of the genotype can also be found in the study of macroevolution. Evolutionary acceleration or deceleration may be displayed by all or at least most members of a new higher taxon. A well-known textbook case is evolution among the lungfishes, as worked out by Westoll (1949). Almost all the anatomical reconstruction in this new class of fishes took place in the earliest stages (during some 25 million years), while almost no change has taken place in the subsequent 200 million years. Such a drastic difference between the rates of evolutionary change in young and mature higher taxa is virtually the rule. Bats diverged from insectivores within a few million years, but their morphology has hardly changed in the last 40 million years. The fossil record (though inadequate) suggests that the same is true for the anatomical reconstruction during the origin of birds and whales. Exceptions are much in the minority.

THE ROLE OF SOMATIC PROGRAMS IN EVOLUTION
The traditional formula according to which development is programmed by the genotype implies far too direct a pathway from the gene to the endpoint of its action in the phenotype. Embryologists since Kleinenberg (1886) have known that genes might effect the production of an embryonic structure which subsequently serves as part of the program for the later stages of development. That not all developmental programs are purely genetic, and that there are also somatic programs, is best made clear by the example of programs controlling behavior. When, for example, a male bird performs a certain courtship display, this action is not programmed directly by the genotype but rather by a secondary program laid down in the central nervous system

during ontogeny. And it is this secondary—somatic—program that actually controls the behavior.

That somatic programs exist is probably not of major consequence for the evolution of behavior. The existence of somatic programs, however, may be of high significance in morphological evolution. It may help to explain many puzzling phenomena of both ontogeny and evolution. For instance, it may explain most instances of recapitulation. When an embryonic structure of a species' ancestors is maintained in ontogeny even though it no longer seems to be of functional value (for instance, the gill arches of the mammalian embryos), such an embryonic structure may have been preserved by natural selection because it serves as the somatic program for the subsequent ontogenetic stages. The existence of somatic programs imposes important constraints on evolution. Much development in the higher taxa seems to be constrained by such somatic programs, which seem to be highly resistant to evolutionary change. This statement is, of course, at the present time only words. However, recent research in molecular ontogeny has made promising beginnings. Progress is bound to be slow, because development involves highly complex interactions between different domains of the genotype and different somatic programs.

This subject area is of particular interest to the student of macroevolution. It is here that the connection will be made between the genetics of individuals and populations on the one hand and the major macroevolutionary processes and events on the other. Research will have to pursue simultaneously reductionist approaches (that is, the study of the action of individual genes) and also holistic ones (that is, the study of domains of the genotype and of whole somatic programs).

Darwinism Today

The basic theory of evolution has been confirmed so completely that modern biologists consider evolution simply a fact. How else except by the word "evolution" can we designate the sequence of faunas and floras in precisely dated geological strata? And evolutionary change is also simply a fact owing to the changes in the content of gene pools from generation to generation. It is as much of a fact as the observation that the earth revolves around the sun rather than the reverse. Evolution is the factual basis on which the other four Darwinian theories rest. For instance, all the phenomena explained by common descent would make no sense if evolution were not a fact.

Darwin's second theory—common descent—also has been gloriously confirmed by all researches since 1859. Everything we have learned about the physiology and chemistry of organisms supports Darwin's daring speculation that "all the organic beings which have ever lived on this earth have descended from some one primordial form, into which life was first breathed" (1859:484). The discovery that the prokaryotes have the same genetic code as the higher organisms was the most decisive confirmation of Darwin's hypothesis. A historical unity in the entire living world cannot help but have a deep meaning for any thinking person and for his feeling toward fellow organisms.

Darwin's great emphasis on the development of diversity as an important component of the evolutionary process — his third theory—was undeservedly neglected during the first third of this century but is again at the forefront of interest, particularly in paleontology and ecology. As far as speciation is concerned—the process that serves as the source of new diversity—Darwin was somewhat confused, having supported geographic speciation on islands and a widespread sympatric speciation on continents. The controversy over speciation is still alive today, but the basic theory that species multiply as well as evolve is uncontested.

Darwin's theory of gradualism, unpalatable even to his close friends Huxley and Galton, has ultimately triumphed decisively and makes more sense the more clearly we realize that evolution is a process involving populations. The only apparent exceptions are the occasional abandonment of sexual reproduction and certain chromosomal processes such as polyploidy. But these processes have not led to any macroevolutionary consequences different from populational evolution. All other speciation is populational, even in the theory of punctuated equilibria.

The greatest triumph of Darwinism is that the theory of natural selection, for eighty years after 1859 a minority opinion, is now the prevailing explanation of evolutionary change. It has achieved this position both by irrefutable proofs and by default, as all the opposing theories were demolished. Darwin took it for granted that a nearly unlimited amount of variation was at all times available to provide raw material for natural selection. He had no idea as to the source of this variation and supported several genetic theories (soft inheritance, pangenesis, blending inheritance) that have since been refuted. Nevertheless, advances in genetics continue to strengthen rather than weaken the theory of natural selection. Darwin was remarkably astute in his conceptualization of selection. He clearly saw (better than A. R. Wallace and most other contemporaries) that there are two kinds of selection, one for general viability leading to survival and the maintenance or improvement of adaptedness, and this he called "natural selection," and another that leads to greater reproductive success, and this he called "sexual selection."

Where evolutionists today differ from Darwin is almost entirely on matters of emphasis. While Darwin was fully aware of the probabilistic nature of selection, the modern evolutionist emphasizes this even more. The modern evolutionist realizes how great a role chance plays in evolution. Darwin never said, "Selection can do anything"; neither do we. On the contrary, there are powerful constraints on selection. And selection is, for various reasons, appallingly often unable to prevent extinction.

One hundred and thirty years of unsuccessful refutations have resulted in an immense strengthening of Darwinism. Controversies within evolutionary biology about such matters as the occurrence of sympatric speciation, the existence or not of cohesive domains within the genotype, the relative frequency of complete stasis in species, the rate of speciation, the significance of neutral allele replacement, and whatever else all take place within the framework of Darwinism. The basic Darwinian principles are more firmly established than ever.

References

Glossary

Acknowledgments

Index

References

Adams, M. 1980. Sergei Chetverikov, the Koltsov Institute, and the evolutionary synthesis. In E. Mayr and W. B. Provine, *The Evolutionary Synthesis*, pp. 242–278. Cambridge: Harvard University Press.

Alexander, R. D. 1979. *Darwinism and Human Affairs*. Seattle: University of Washington Press.

Ayala, F. J. 1982. Beyond Darwinism? The challenge of macroevolution to the synthetic theory of evolution. *Philosophy of Science Association (PSA)* 2:275–291.

———— 1983. Microevolution and macroevolution. In D. S. Bendall, ed., *Evolution from Molecules to Men*, pp. 387–402. Cambridge: Cambridge University Press.

Barlow, N., ed. 1963. Darwin's ornithological notes. *Bull. Brit. Mus. (Nat. Hist.)*, Hist. Ser., 2:201–278.

Barrett, P. H., P. J. Gautrey, S. Herbert, D. Kohn, and S. Smith. 1987. *Charles Darwin's Notebooks, 1836–1844*. Ithaca, NY: Cornell University Press.

Barzun, J. 1958. *Darwin, Marx and Wagner: Critique of a Heritage*. 2nd. ed. Garden City, NY: Anchor.

Bateson, W. 1894. *Materials for the Study of Variation*. London: Macmillan.

Beckner, M. 1969. Function and teleology. *J. Hist. Biol.* 2:151–164.

Beneden, E. van. 1883. Recherches sur la maturation de l'oeuf et la fécondation. *Arch. Biol.* 4:265–632.

Bergson, H. 1907. *Evolution Créative*. Paris: Alcan.

Blacher, L. I. 1982. *The Problem of the Inheritance of Acquired Characters: A History of A Priori and Empirical Methods Used to Find a Solution*. New Delhi: Amerind Publishing Co. Eng. trans. ed. F. B. Churchill. Russian original, Moscow: Nauka, 1971.

Bonnet, C. 1781. *Contemplation de la nature*. New edition.

Bowler, P. J. 1976. *Fossils and Progress*. New York: Science History Publications.

—— 1977. Darwinism and the argument from design: suggestions for a reevaluation. *J. Hist. Biol.* 10:29–43.

—— 1979. Theodor Eimer and orthogenesis: evolution by 'definitely directed variation.' *J. Hist. Med. Allied Sci.* 34:40–73.

—— 1983. *The Eclipse of Darwinism.* Baltimore, MD: Johns Hopkins University Press.

—— 1988. *The Non-Darwinian Revolution.* Baltimore: Johns Hopkins University Press.

Buch, L. von. 1825. *Physicalische Beschreibung der Canarischen Inseln,* pp. 132–133. Berlin: Kgl. Akad. Wiss.

Buffon, G. L. 1749. *Histoire Naturelle.* Paris: Imprimerie Royale.

Burian, R. M. 1989. The influence of the evolutionary paradigms. In M. K. Hecht, ed., *Evolutionary Biology at the Crossroads,* pp. 149–166. Flushing, NY: Queens College Press.

Churchill, F. B. 1968. August Weismann and a break from tradition. *J. Hist. Biol.* 1:91–112.

—— 1979. Sex and the single organism: biological theories of sexuality in mid-nineteenth century. *Stud. Hist. Biol.* 3:139–177.

—— 1985. Weismann's continuity of the germ plasm in historical perspective. *Freiburger Universitäts-blätter* 87/88:107–124.

Darwin, C. 1839. *Journal of Researches.* London: John Murray.

—— 1842. Sketch. In F. Darwin, 1909, p. 46.

—— 1844. Essay. In F. Darwin, 1909.

—— 1859. *On the Origin of Species by Means of Natural Selection or the Preservation of Favored Races in the Struggle for Life.* London: Murray. [1964 facsimile ed.]

—— 1868. *The Variation of Animals and Plants under Domestication.* London: Murray.

—— 1958. *The Autobiography of Charles Darwin.* Ed. Nora Barlow. London: Collins.

—— 1975. *Natural Selection.* Ed. R. C. Stauffer. Cambridge: Cambridge University Press.

—— 1988. *The Correspondence of Charles Darwin.* Ed. F. S. Smith. Vol. 4. Cambridge: Cambridge University Press.

—— and A. R. Wallace. 1858. [evolution by natural selection] *Linn. Soc. London* [also G. de Beer, ed. Cambridge: Cambridge University Press, 1958].

Darwin, F. 1888. *The Life and Letters of Charles Darwin.* 3 vols. London: Murray. Rptd. New York: Johnson Reprint Corp., 1969.

—— 1909. *The Foundations of the Origin of Species.* Cambridge: Cambridge University Press.

—— and A. C. Seward, eds. 1903. *More Letters of Charles Darwin.* 2 vols. London: Murray.

De Beer, G. R. 1961. The origins of Darwin's ideas on evolution and natural selection. *Proc. Roy. Soc. London* 155:321–378.

de Candolle, A. S. 1820. *Essai Elementaire de Géographie Botanique.*

Derham, W. 1713. *Physico-Theology, or, Demonstration of the Being and Attributes of God from His Works of Creation.* London.

Desmond, A. J. 1982. *Archetypes and Ancestors: Paleontology in Victorian London, 1850–1875.* London: Blond and Briggs.

de Vries, H. 1901. *Die Mutationstheorie. Versuche und Beobachtungen über die Entstehung der Arten im Pflanzenreich.* Vol. I, *Die Entstehung der Arten durch Mutation.* Leipzig: Veit. Eng. trans. J. B. Farmer and A. D. Darbishire. Chicago: Open Court Publishing Co., 1909–10.

Dobzhansky, Th. 1937. *Genetics and the Origin of Species.* 1st. ed. New York: Columbia University Press.

Eldredge, N. 1985. *Time for Change.* New York: Simon and Schuster.

——— and S. J. Gould. 1972. Punctuated equilibria: an alternative to phyletic gradualism. In T. J. M. Schopf, ed., *Models in Paleobiology,* pp. 82–115. San Francisco: Freeman, Cooper.

Farley, J. 1982. *Gametes and Spores: Ideas about Sexual Reproduction.* Baltimore, MD: Johns Hopkins University Press.

Freeman, D. 1974. The evolutionary theories of Charles Darwin and Herbert Spencer. *Current Anthropology* 15:211–237.

Futuyma, D. 1983. *Science on Trial: The Case for Evolution.* New York: Pantheon Books.

Ghiselin, M. T. 1969. *The Triumph of the Darwinian Method.* Berkeley: University of California Press.

Gillespie, N. C. 1979. *Charles Darwin and the Problem of Creation.* Chicago: University of Chicago Press.

Goldschmidt, R. 1940. *The Material Basis of Evolution.* New Haven: Yale University Press.

Goodman, M. 1982. Molecular evolution above the species level. *Syst. Zool.* 31:376–399.

Gordon, S. 1989. Darwin and political economy: the connection reconsidered. *J. Hist. Biol.* 22:437–459.

Gould, S. J. 1977. *Ontogeny and Phylogeny.* Cambridge, MA: Harvard University Press.

——— 1980. Is a new and general theory of evolution emerging? *Paleobiology* 6:119–130.

Grant, V. 1983. The synthetic theory strikes back. *Biol. Zentralbl.* 102:149–158.

Gray, A. 1876. *Darwiniana.* New York: D. Appleton. Rpt. ed. H. Dupree, Cambridge, MA: Harvard University Press, 1963.

Greene, J. C. 1986. The history of ideas revisited. *Revue de Synthèse* (4)3:201–227.

Gruber, H. E. 1974. *Darwin on Man*. In H. E. Gruber and P. H. Barrett, *Darwin on Man: A Psychological Study of Scientific Creativity. Together with Darwin's Early and Unpublished Notebooks*, pp. 1–257. New York: Dutton, 1974.

———— 1981. *Darwin on Man*. 2nd ed. Chicago: University of Chicago Press.

Gutmann, W. F., and K. Bonik. 1981. *Kritische Evolutions-theorie. Ein Beitrag zur Überwindung altdarwinistischer Dogmen*. Hildesheim: Gerstenberg.

Haeckel, E. 1866. *Generelle Morphologie der Organismen*. 2 vols. Berlin: Georg Reimer.

Haldane, J. B. S. 1932. *The Causes of Evolution*. London: Longmans, Green and Co.

———— 1959. An Indian perspective of Darwin. *Cent. Rev. Arts Sci., Mich. State Univ.* 3:357.

Harris, H. 1966. Enzyme polymorphism in man. *Proc. Roy. Soc.*, B 164:298–316.

Hartmann, E. v. 1872. *Das Unbewusste vom Standpunkt der Physiologie und Deszendenzlehre*. Berlin: Carl Duncker.

Herbert, S. 1971. Darwin, Malthus, and selection. *J. Hist. Biol.* 4:209–217.

Herder, J. G. 1784. *Ideen zur Philosophie der Geschichte der Menschheit*, II, 3:89.

Herschel, J. F. W. 1830. *Preliminary Discourse on the Study of Natural History*. London: Longmans.

Hessen, B. 1931. The social and economic roots of Newton's Principia, pp. 147–212 in N. I. Bukharin et al., *Science at the Crossroads*. London: Cass.

Himmelfarb, G. 1959. *Darwin and the Darwinian Revolution*. London: Doubleday.

Hodge, M. J. S. 1982. Darwin and the laws of the animate past of the terrestrial system (1835–1837). *Stud. Hist. Biol.* 7:1–106.

———— and D. Kohn. 1985. The immediate origins of natural selection. In D. Kohn, ed., *The Darwinian Heritage*, pp. 185–206. Princeton: Princeton University Press.

Hoffman, A. 1989. *Arguments on Evolution: A Paleontologist's Perspective*. New York: Oxford University Press.

Hubby, J. L., and R. C. Lewontin. 1966. The number of alleles at different loci in *Drosophila pseudoobscura*. *Genetics* 54:577–594.

Hull, D. L. 1973. *Darwin and His Critics: The Reception of Darwin's Theory of Evolution by the Scientific Community*. Cambridge, MA: Harvard University Press.

———— 1983. Exemplars and scientific change. *Philosophy of Science Association (PSA)* 2:479–503.

———— 1985. Darwinism as a historical entity: historiographic proposal. In D. Kohn, ed., *The Darwinian Heritage*, pp. 773–812. Princeton: Princeton University Press.

Hume, D. 1779. *Dialogues Concerning Natural Religion*. London.

Huxley, A. 1982. [Address of the President.] *Proc. Roy. Soc. London A* 379:ix–xvii.

Huxley, J. 1942. *Evolution: The Modern Synthesis*. London: Allen and Unwin.

Huxley, T. H. 1864. *Nat. Hist. Rev.*, October, p. 567.

———— 1900. *Life and Letters of Thomas Henry Huxley, by His Son Leonard Huxley*. 2 vols. London: Macmillan.

Jacob, F. 1977. Evolution and tinkering. *Science* 196:1161–1166.

Jenkin, F. 1867. The origin of species. *North Brit. Rev.* 42:149–171.

Kellogg, V. L. 1907. *Darwinism To-Day*. New York: Henry Holt.

Kimura, M. 1968. Evolutionary rate at the molecular level. *Nature* 217:624–626.

King, J. L., and T. H. Jukes. 1969. Non-Darwinian evolution. *Science* 164:788–798.

Kitcher, P. 1982. *Abusing Science: The Case against Creationism*. Cambridge, MA: MIT Press.

———— 1985. Darwin's achievements. In N. Rescher, ed., *Reason and Rationality in Natural Science*, pp. 127–189. New York: University Press of America.

Kleinenberg, N. 1886. Über die Entwicklung durch Substitution von Organen. In A. v. Kölliker and E. Ehlers, eds., *Zeitschrift für Wissenschaftliche Zoologie*. Leipzig: Wilhelm Engelmann.

Kohn, D. 1975. *Charles Darwin's Path to Natural Selection*. Ph.d. diss., University of Massachusetts.

———— 1980. Theories to work by: rejected theories, reproduction, and Darwin's path to natural selection. *Stud. Hist. Biol.* 4:67–170.

————, ed. 1985. *The Darwinian Heritage*. Princeton: Princeton University Press.

———— 1989. Darwin's ambiguity: the secularization of biological meaning. *Brit. J. Hist. Sci.* 22:215–239.

Kölliker, A. v. 1864. *Ueber die Darwinische Schöpfungstheorie*. Leipzig.

Kottler, M. 1978. Charles Darwin's biological species concept and theory of geographic speciation. *Amer. Sci.* 35:275–297.

Lamarck, J.-B. 1809. *Philosophie Zoologique*. Paris.

Laporte, L. F. 1983. Simpson's *Tempo and Mode in Evolution* revisited. *Proc. Amer. Phil. Soc.* 127:365–417.

Lennox, J. G. 1983. Robert Boyle's defense of teleological inference in experimental science. *Isis* 74:38–52.

Lenoir, T. 1982. *The Strategy of Life: Teleology and Mechanics in the 19th Century.* Dordrecht: Reidel.

Levinton, J. 1988. *Genetics, Paleontology, and Macroevolution.* Cambridge: Cambridge University Press.

Lewontin, R. 1983. The organism as the subject and object of evolution. *Scientia* 118:63–82.

Limoges, C. 1970. *La sélection naturelle.* Paris: Presses Universitaires de France.

Linnaeus, C. 1781. *Politica Naturae [Amoen. Academicae],* pp. 131–32, trans. F. J. Brand. London.

Lovejoy, A. O. 1936. *The Great Chain of Being.* Cambridge, MA: Harvard University Press.

Löw, R. 1980. *Philosophie des Lebendigen.* Frankfurt: Suhrkamp.

Lyell, C. 1830–1833. *Principles of Geology.* 3 vols. 1st ed. London: Murray.

——— 1835. *Principles of Geology.* 4th ed. London: Murray.

Malthus, T. R. 1798. *An Essay on the Principle of Population, as It Affects the Future Improvement of Society.* London: J. Johnson. [Darwin actually read the 6th ed., London: Murray, 1826.]

Maynard Smith, J. 1983. Current controversies in evolutionary biology. In M. Greene, ed., *Dimensions of Darwinism.* Cambridge: Cambridge University Press, 1983.

———, ed. 1982. *Evolution Now: A Century after Darwin.* London: Macmillan.

Mayr, E. 1942. *Systematics and the Origin of Species.* New York: Columbia University Press.

——— 1959a. Agassiz, Darwin, and evolution. *Harvard Library Bull.* 13:165–194.

——— 1959b. Darwin and the evolutionary theory in biology. In B. J. Meggers, *Evolution and Anthropology: A Centennial Appraisal,* pp. 1–10. Washington: Anthropological Society of Washington.

——— 1963. *Animal Species and Evolution.* Cambridge, MA: Harvard University Press.

——— 1982. *The Growth of Biological Thought.* Cambridge, MA: Harvard University Press.

——— 1983a. Comments on David Hull's paper on exemplars and type specimens. *Philosophy of Science Association (PSA)* 2:504–511.

——— 1983b. How to carry out the adaptationist program? *Amer. Nat.* 121:324–334.

——— 1984. The triumph of the evolutionary synthesis. *Times Literary Supplement* no. 4 (257), November 2, pp. 1261–1262.

——— 1988. *Toward a New Philosophy of Biology.* Cambridge, MA: Harvard University Press.

———— 1989. Attaching names to objects. In M. Ruse, ed., *What the Philosophy of Biology Is*, pp. 235–243. Dordrecht: Kluwer Academic.

———— 1990. The myth of the non-Darwinian revolution. (Review of Peter J. Bowler, *The Non-Darwinian Revolution, 1988.*) *Biology and Philosophy* 5:85–92.

———— and W. B. Provine, eds. 1980. *The Evolutionary Synthesis: Perspectives on the Unification of Biology.* Cambridge, MA: Harvard University Press.

Mill, J. S. 1843. *A System of Logic.* London: Longmans, Green.

Monod, J. 1970. *Le hasard et la necessité.* Paris: Seuil.

Montague, A., ed. 1983. *Science and Creationism.* Oxford: Oxford University Press.

Moore, J. R. 1979. *The Post-Darwinian Controversies.* Cambridge: Cambridge University Press.

———— 1989. Of love and death: why Darwin gave up Christianity. In J. R. Moore, ed., *History, Humanity, and Evolution*, pp. 195–229. Cambridge: Cambridge University Press.

Morgan, T. H. 1910. Chance or purpose in the origin and evolution of adaptation. *Science* 21:201–210.

Muller, H. J. 1940. Bearings of the 'Drosophila' work on systematics. In J. Huxley, ed., *The New Systematics*, pp. 185–268. Oxford: Clarendon Press.

Nagel, E. 1961. The structure of teleological explanations. In *The Structure of Science*. New York: Harcourt, Brace and World.

Nevo, E. 1983. Adaptive significance of protein variation. In G. S. Oxford and D. Rollinson, eds., *Protein Polymorphism: Adaptive and Taxonomic Significance*, pp. 239–282. Systematic Association Special Volume No. 24. New York: Academic Press.

Newell, N. D. 1982. *Creation and Evolution: Myth or Reality?* New York: Columbia University Press.

Osborn, H. F. 1894. *From the Greeks to Darwin: An Outline of the Development of the Evolution Idea.* New York: Columbia University Press.

Ospovat, D. 1981. *The Development of Darwin's Theory: Natural History, Natural Theology, and Natural Selection, 1838–1859.* Cambridge: Cambridge University Press.

Otte, D., and J. A. Endler. 1989. *Speciation and Its Consequences.* Sunderland, MA: Sinauer.

Provine, W. B. 1971. *The Origins of Theoretical Population Genetics.* Chicago: University of Chicago Press.

Ray, J. 1691. *The Wisdom of God Manifested in the Works of the Creation.*

Recker, D. A. 1987. Causal efficacy: the structure of Darwin's argument strategy in the *Origin of Species. Phil. Sci.* 54:147–175.

———— 1990. There's more than one way to recognize a Darwinian: Lyell's Darwinism. *Phil. Sci.* 57 (3):459–478.

Rensch, B. 1947. *Neuere Probleme der Abstammungslehre*. Stuttgart: Enke.

Robertson, M. 1985. Mice, mating types, and molecular mechanisms of morphogenesis. *Nature* 318:12–13.

Romanes, G. J. 1896. *Darwin and after Darwin*. 3 vols. Chicago: Open Court.

Ruse, M. 1975a. Charles Darwin and artificial selection. *J. Hist. Ideas* 36:339–350.

———— 1975b. Darwin's debt to philosophy. *Stud. Hist. Phil. Sci.* 6:159–181.

———— 1979. *The Darwinian Revolution*. Chicago: University of Chicago Press.

———— 1982. *Darwinism Defended: A Guide to the Evolution Controversies*. Reading, MA: Addison-Wesley.

Sachs, J. 1894. Mechanomorphosen und Phylogenie. *Flora* 78:215–243.

Sapp, J. 1987. *Beyond the Gene*. New York: Oxford University Press.

Schindewolf, O. H. 1950. *Grundfragen der Paläontologie*. Stuttgart: Schweizerbart.

Schleiden, M. J. 1842. *Grundzüge der wissenschaftlichen Botanik*. Leipzig: Wilhelm Engelmann.

Sigwart, C. 1881. Der Kampf gegen den Zweck. *Kleine Schriften* 2:24–67.

Simpson, G. G. 1944. *Tempo and Mode of Evolution*. New York: Columbia University Press.

———— 1949. *The Meaning of Evolution*. New Haven: Yale University Press.

———— 1953. *The Major Features of Evolution*. New York: Columbia University Press.

———— 1961. *Principles of Animal Taxonomy*. New York: Columbia University Press.

———— 1964. *This View of Life*. New York: Harcourt, Brace, and World.

———— 1974. The concept of progress in organic evolution. *Social Research*, pp. 28–51.

Smith, S. 1960. The origin of "The Origin." *Advancement of Science* 64:391–401.

Stanley, S. M. 1981. *The New Evolutionary Timetable*. New York: Basic Books.

Stebbins, G. L. 1950. *Variation and Evolution in Plants*. New York: Columbia University Press.

———— and F. J. Ayala. 1981. Is a new evolutionary synthesis necessary? *Science* 213:967–971.

Sulloway, F. J. 1979. Geographic isolation in Darwin's thinking: the vicissitudes of a crucial idea. *Stud. Hist. Biol.* 3:23–65.

———— 1982a. Darwin and his finches: the evolution of a legend. *J Hist. Biol.* 15:1-53.

———— 1982b. Darwin's conversion: the *Beagle* voyage and its aftermath. *J. Hist. Biol.* 15:327-398.

———— 1984. Darwin and the Galapagos. *Biol. J. Linn. Soc.* 21:29-59.

Toulmin, S. 1972. *Human Understanding.* Princeton: Princeton University Press.

———— 1982. Darwin und die Evolution der Wissenschaften. *Dialektik* 5:68-78.

Trivers, R. 1985. *Social Evolution.* Menlo Park, CA: Benjamin/Cummings.

Tuomi, J. 1981. Structure and dynamics of Darwinian evolutionary theory. *Syst. Zool.* 30:22-31.

Waddington, C. F. 1960. *The Ethical Animal.* London: Allen and Unwin.

Wagner, M. 1841. *Reisen in der Regentschaft Algier in den Jahren 1836, 1837 und 1838.* Leipzig: Leopold Voss.

———— 1868. *Die Darwinische Theorie und das Migrationsgesetz der Organismen.* Leipzig: Dunker und Humblot.

———— 1889. *Die Entstehung der Arten durch räumliche Sonderung.* Basel: Benno Schwalbe.

Wallace, A. R. 1855. On the law which has regulated the introduction of new species. *Ann. Mag. Nat. Hist.* 16:184-196. Rptd. in *Natural Selection and Tropical Nature,* pp. 3-19. London: Macmillan, 1891.

———— 1889. *Darwinism.* London: Macmillan.

Weismann, A. 1868. *Über die Berechtigung der Darwinschen Theorie.* Leipzig: Engelmann.

———— 1872. *Über den Einfluss der Isolirung auf die Artbildung.* Leipzig: Engelmann.

———— 1882. *Studies in the Theory of Descent.* Trans. R. Mendola, London: Sampson, Low, et al. (Eng. transl. of Weismann 1875, 1876.)

———— 1883. *Über die Vererbung.* Jena: G. Fischer.

———— 1886. *Die Bedeutung der sexuellen Fortpflanzung für die Selektionstheorie.* Jena: G. Fischer.

———— 1891. *Amphimixis; oder, Die Vermischung der Individuen.* Jena: G. Fischer.

———— 1892. *Das Keimplasma: Eine Theorie der Vererbung.* Jena: G. Fischer. Engl. ed., 1893.

———— 1893. *Die Allmacht der Naturzüchtung: Eine Erwiderung an Herbert Spencer.* Jena: G. Fischer.

———— 1896. Über Germinal-Selection: Eine Quelle bestimmt gerichteter Variation. Jena: G. Fischer.

———— 1904. *Vorträge über Deszendenztheorie.* 2 vols. 2nd ed. Jena: G. Fischer. 1st ed., 1902; 3rd. ed., 1910.

———— 1909. *Die Selektionstheorie: Eine Untersuchung.* Jena: G. Fischer. ("The Selection Theory." In A. C. Seward, ed., *Darwin and Modern Science,* pp. 18–65. Cambridge: Cambridge University Press, 1909.)

Westoll, S. 1949. On the evolution of the Dipnoi. In G. Jepsen, E. Mayr, and G. G. Simpson, eds., *Genetics, Paleontology, and Evolution,* pp. 121–184. Princeton, NJ: Princeton University Press.

Whewell, W. 1840. *Philosophy of the Inductive Sciences.* London: Parker.

White, M. J. D. 1978. *Modes of Speciation.* San Francisco: Freeman.

———— 1981. Tales of long ago. *Paleobiology* 7:287–291.

Willis, J. C. 1940. *The Course of Evolution by Differentiation or Divergent Evolution Rather Than by Selection.* Cambridge: Cambridge University Press.

Willmann, R. 1985. *Die Art in Raum und Zeit.* Berlin und Hamburg: Parey.

Wilson, A. C., V. M. Sarich, and L. R. Maxson. 1974. The importance of gene rearrangement in evolution: evidence from studies on rates of chromosomal, protein, and anatomical evolution. *Proc. Nat. Acad. Sci.* 71:3028–3030.

Wilson, E. O. 1975. *Sociobiology: The New Synthesis.* Cambridge, MA: Harvard University Press.

Young, W. 1985. *Fallacies of Creationism.* Calgary, Alberta: Detselig Enterprises Ltd.

Glossary

Acquired characters Those characteristics of an organism's appearance (phenotype) that result from environmental influences rather than inheritance

Adaptationist program The research endeavor to discover the adaptive significance of structures and behaviors

Adaptedness The suitability of a structure or an organism for its environment or lifestyle, as a result of past selection

Additive inheritance Stress on the independent effect of individual genes, with disregard for the mutual (epistatic) interaction of genes

Allele Any of the alternative variants of a gene

Allopatric speciation See Geographical speciation

Altruism Behavior that benefits another organism at a cost to the actor, where cost and benefit are defined in terms of reproductive success

Artificial selection Selection of breeding stock by an animal or plant breeder

Base pair A pair of hydrogen-bonded nitrogenous bases (one purine and one pyrimidine) that connect the two strands of the DNA double helix. In DNA, adenine pairs only with thymine, and guanine pairs only with cytosine. This complementarity is the key to the self-replication and information-transmitting capabilities of DNA

Bauplan The basic structural characteristics (morphotype) of a given type of organism, such as the *Bauplan* of the vertebrates, or of the birds, or of the insects

Biological species concept Defines a species as a reproductively isolated aggregate of populations which can interbreed because they share the same isolating mechanisms. See Evolutionary species concept; Nominalist species concept; Typological species concept

Biota   Fauna and flora

Blending inheritance   A now-disproven theory that the genetic determinants of the parents fuse into a uniform substance during the fertilization of the egg. See Particulate inheritance

Catastrophism   A theory that catastrophic events in the history of the earth have resulted in the partial or complete extinction of the biota

Causations, proximate   See Proximate causations; Ultimate causations

Central dogma   The assertion that the information contained in proteins cannot be translated into nucleic acids

Character   A component of the phenotype

Character divergence   The name given by Darwin to the differences developing in two (or more) related species in their area of coexistence, owing to the selective effects of competition

Chromosome   One of the threadlike structures in the nucleus of the cell, consisting of DNA and associated protein. See DNA; Gene

Common descent   The derivation of certain species or higher taxa from a common ancestor

Cosmic teleology   The belief that the universe as a whole or some of its changes are directed toward a final objective or goal, such as greater perfection

Creationism   A belief in the literal truth of the story of creation as recorded in the Book of Genesis

Deoxyribonucleic acid   See DNA

Deism   A belief in a supreme being who rules the world through divinely ordained laws rather than through direct intervention

Determinism   A theory that the outcome of any process is strictly predetermined by definite causes and natural laws and is therefore predictable

Diploid   Having a double set of chromosomes. See Haploid

DNA (deoxyribonucleic acid)   The molecule that carries the genetic information (genes) in all organisms except the RNA viruses. It consists of two long polysugar-phosphate strands connected by base pairs and twisted in a double helix. See Chromosome; Gene

Epigenesis Development from an unformed basic material, as opposed to preformation. See Preformation

Essentialism The belief, going back to Plato, that the changing variety of nature can be sorted into a limited number of classes, each of which can be defined by its essence. Variation is simply the manifestation of imperfect representation of these constant essences. Also referred to as typological thinking

Eukaryotes Organisms whose cells have a nucleus with discrete chromosomes as well as such cellular organelles as mitochrondria and chloroplasts

Evolutionary causations See Ultimate causations

Evolutionary species concept Defines a species as a phyletic lineage evolving separately from others, with its own evolutionary tendencies and historical fate. See Biological species concept; Nominalistic species concept; Typological species concept

Evolutionary synthesis A somewhat modified Darwinian paradigm which includes a refutation of transformational evolution, saltationism, and orthogenesis, while placing great emphasis on natural selection, adaptation, and the study of diversity (origin of species and higher taxa). This paradigm was worked out in the 1930s and 1940s by a group of "architects" of the evolutionary synthesis, which included Dobzhansky, Mayr, Rensch, Simpson, Stebbins, and Timofeeff-Ressovsky. The term is sometimes also used to refer to the time period when the paradigm was being constructed, and the paradigm is sometimes referred to as the modern synthesis

Exons Sequences of base pairs in a gene that participate in the coding of the peptides. See Introns

Female choice A theory in sexual selection which states that it is often the female that selects one of several available males for mating, rather than the other way around

Finalism A belief in an inherent trend in the natural world toward some preordained final goal or purpose, such as the attainment of perfection. See Teleology

Fisherism The set of evolutionary theories held or proposed by R. A. Fisher which stressed the power of selection, the gene as the unit of selection, the particulateness of inheritance, and the almost exclusive prevalence of additive inheritance. Evolution for Fisher was a change in gene frequencies within a population

Fitness The relative ability of an organism to survive and transmit its genes to the gene pool of the next generation

Gamete A germ cell (egg or sperm) carrying half of the organism's full set of chromosomes; especially a mature germ cell capable of participating in fertilization. See Genetic recombination; Meiosis

Gene In classical genetics, a unit of inheritance, transmitted from generation to generation by an ovum or sperm, which controls some characteristic of an individual or some aspect of the individual's development. In molecular biology, a sequence of base pairs in a DNA molecule that contains information for the construction of one protein molecule (peptide). See Base pair; Chromosome; DNA

Gene pool The totality of the genes of a population or a species

Genetic code The code that determines which triplets of base pairs are translated into which amino acids

Genetic drift Changes in the gene content of a population owing to "chance," that is, to stochastic processes

Genetic program The information coded in an organism's DNA

Genetic recombination The reshuffling of an organism's genes during the production of germ cells, through crossing over of sections of the organism's maternal and paternal chromosomes. It occurs during meiosis, just prior to cell division and the chromosomes' independent assortment during the reduction division. Genetic recombination assures that the chromosomes carried by an organism's eggs or sperm are not identical to the chromosomes the organism inherited from either of its parents. No two chromosomes of any of the eggs or sperm are likely to be identical to one another. Genetic recombination, as now recognized, is responsible for a large share of the variation upon which natural selection acts. See Meiosis

Genome The totality of genes carried by a single gamete

Genotype The genetic constitution of an individual, especially as distinguished from its physical appearance. See Phenotype

Geographical speciation Speciation that occurs while populations are geographically isolated; also known as allopatric speciation

Germ cell A cell whose principal function is reproduction; an egg or sperm cell

Germ plasm An outmoded term which referred to the genetic material in the germ cells, as distinguished from the "soma," that is, the phenotype

Gradualism A theory that evolution progresses by the gradual modification of populations, and not by the sudden origin of new types (saltations). See Saltational evolution

Great chain of being See Scala naturae

Haploid Having the number of chromosomes of a normal germ cell, which equals half the number in a somatic cell

Hard inheritance A theory that the genetic material is constant ("hard") and cannot be affected by lifestyle or the environment. None of the changes in the phenotype of an organism during its lifespan can be passed on to its offspring. In the terminology of molecular biology, hard inheritance is the theory that information in the proteins cannot be conveyed to the nucleic acids in DNA; also known as the "central dogma." See Blending inheritance; Inheritance of acquired characters; Particulate inheritance; Soft inheritance

Heterozygosity The occurrence of two different variants (alleles) of a gene at the corresponding loci on two homologous chromosomes. See Homozygosity

Homeobox A definite sequence of genes controlling a step in development, particularly in metameric organisms

Homozygosity The occurrence of two identical alleles at the corresponding loci on two homologous chromosomes. See Heterozygosity

Horizontal evolution The simultaneous evolution of geographically distributed populations; geographic variation

Hybrid The product of the interbreeding of different kinds of organisms, ordinarily of different species

Incipient species A population in the process of evolving into a separate species

Inclusive fitness The sum of an individual's own fitness plus all its influence on fitness in its relatives other than direct descendants

Infusorians One-celled organisms; protists

Inheritance of acquired characters A now-disproven theory that changes in an organism's phenotype which result from factors in its environment can be passed on to offspring through the organism's genetic material

Introns Noncoding sequences of base pairs that are eliminated prior to the translation of the nucleic acids into proteins (peptides)

Isolating mechanisms Biological or behavioral properties of individuals which prevent the interbreeding of populations that coexist in the same area

Kin selection Selection for the shared components of the genotype in individuals related by common descent

Lamarckism A belief in the gradual change of species over time to a "higher" (meaning better adapted and more complex) level through the inheritance of acquired characters

Macroevolution Evolution above the species level; the evolution of higher taxa and the production of evolutionary novelties such as new structures. See Microevolution

Meckel-Serres Law. A law which states that there are parallels between the stages in ontogeny and the phylogenetic series

Meiosis A special series of cell divisions during the production of the gametes in which the number of chromosomes is reduced. This process precedes the production of gametes in animals and spores in plants. It normally involves crossing over and an independent assortment of homologous chromosomes. See Genetic recombination

Mendelism In genetics, a stress on hard inheritance and the particulateness of the units of inheritance, as concluded from Mendel's laws. The early Mendelians also believed in drastic mutations (saltational evolution) and they minimized the importance of natural selection

Metameric Consisting of a seriation of parts

Microevolution Evolution at and below the species level. See Macroevolution

Monotypic Used to describe a taxon that contains only one taxon of the next lower category, such as a genus that contains only one species. See Polytypic

Morphology The science of form and structure in animals and plants

Mosaic evolution Evolution which proceeds at different rates for different structures or for other components of the phenotype and genotype

Mutation In molecular biology, a change in the genotype. If the change occurs in the DNA of a somatic cell, the mutation may cause a change in the organism's phenotype (leading, for example, to cancer) but will not affect the organism's offspring; only mutations in the germ cells can cause heritable changes in the offspring. For the Mendelians, mutation was thought to be the process that produces new species

Natural theology The study of nature to document evidence for the power and wisdom of the Creator in the design of His world

Natural selection The nonrandom survival and reproductive success of a small percentage of the individuals of a population owing to their possession of, at that moment, characters which enhance their ability to survive and reproduce. See Artificial selection; Sexual selection

Neo-Darwinism A theory of evolution developed by August Weismann in the late nineteenth century which consisted of Darwinism without the inheritance of acquired characters and with a strong emphasis on natural selection

Neutral evolution The occurrence and accumulation of heritable mutations which do not affect the fitness of the individual or its offspring

Nominalistic species concept Defines a species as the arbitrary bracketing of individuals under a species name. See Biological species concept; Evolutionary species concept; Typological species concept

Niche The constellation of environmental factors into which a species fits or which it requires for its survival and reproductive success

Nondimensional Without a geographical or time dimension. A nondimensional species is a species as encountered at a particular locality and time, without reference to its geographical relation to other populations or to its evolutionary history

Ontogeny The development of the individual from the fertilized egg (zygote) to adulthood

Orthogenesis The evolution of phyletic lineages along a predetermined linear pathway, not through natural selection

Paleontology The study of fossils and ancient life forms

Pangenesis A theory adopted by Charles Darwin to provide a mechanism by which an inheritance of acquired characters might be explained. It involved the transfer of granules (gemmules) from the body to the gonads and germ cells

Parapatric speciation The progressive divergence of two neighboring populations, while meeting in a contact zone, until they have become two different species

Particulate inheritance A theory developed in the late nineteenth century and adopted by the Mendelians that inheritance is affected by genetic units which do not fuse or blend in the offspring but remain discrete. See Blending inheritance; Hard inheritance

Peripatric speciation Speciation by budding; that is, the origin of new species through the modification of peripherally isolated founder populations

Phenon (pl. phena) A sample of phenotypically similar individuals; a subdivision of a population or species characterized by phenotypic similarity

Phenotype The totality of the observable characteristics of an individual, resulting from interactions between the environment an organism encounters and the genotype it inherited. See Genotype

Phylogeny The origin and subsequent evolution of the higher taxa; the history of the evolutionary lineages

Physicalism A philosophy of science widely held in the seventeenth century and later which asserted that all natural processes (including those in living organisms) can be completely and sufficiently described in chemicophysical terms and that any valid statement in any field of science can, in principle, be reduced to an empirically verifiable physical statement. Physicalism originally included an acceptance of essentialism, determinism, and a reliance on universal laws, and it included no reference to time (historical processes). This theory has been largely abandoned, and in other parts considerably modified, owing to developments in biology and modern physics

Physicotheology See Natural theology

Polymorphism The coexistence of several well-defined distinct phenotypes or alleles in a population

Polyploidy An increase, usually doubling, of the number of chromosomes over what is normally found in the somatic cells, due to a doubling of the chromosomes of the nucleus, not followed by a division of the cell

Polytypic Used to describe a taxon that contains more than one taxon of the next lower category, such as a genus that contains several species. See Monotypic

Population thinking A viewpoint which emphasizes the uniqueness of every individual in populations of a sexually reproducing species and therefore the real variability of populations; the opposite of typological thinking. See Essentialism

Population In evolutionary biology, the community of potentially interbreeding individuals, particularly at a given locality

Preformation Development of the embryo from material in which the eventual form of the adult is "preformed," that is, already exists in its essential structures

Prokaryotes Organisms without a structured cell nucleus, such as various kinds of bacteria and the so-called blue-green algae

Punctuated equilibria See Punctuationism

Punctuationism The theory that most evolutionarily important events take place during short bouts of speciation, and that once species are formed they are relatively stable, sometimes for very long periods. Also known as speciational evolution

Quinarianism An obsolete theory of classification according to which each category contains a circle of five taxa. Each taxon is in contact ("osculates") with a taxon in a different circle, owing to affinity with it

Recapitulation The passing of an individual organism, during its development (ontogeny), through the same stages of development as did its ancestors during phylogeny. See Ontogeny; Phylogeny

Recombination See Genetic recombination

Rectilinear series A series of fossils in a phyletic lineage, seemingly progressing in a linear manner

Reductionism A philosophy which states that all phenomena and laws relating to complex systems (including living ones) can be reduced without residue to those of the physical sciences, and more particularly, to the smallest components. See Physicalism

Saltation In evolutionary theory, the assertion that new types of organisms originate by the sudden origin of a single new individual which becomes the progenitor of this new kind of organism

Saltational evolution Change owing to the sudden origin of a new type, that is, the production of a new kind of individual who gives rise to a new group of organisms. See Transformational evolution; Variational evolution

Scala naturae A linear arrangement of the forms of life from the lowest, nearly inanimate, to the most perfect

Selectionism The theory that adaptive changes in evolution are the result of natural selection

Sexual selection The increased reproductive success of a small percentage of the individuals of a population owing to their possession of, at that moment, characters which enhance either their ability to compete with members of the same sex or their attractiveness to members of the opposite sex. See Natural selection

Sibling species   Morphologically similar or identical populations that are reproductively isolated

Sociobiology   The systematic study of the biological basis of all social behavior

Soma   The body; the phenotype, in contrast with the genotype

Somatic program   A structure or stage in development providing information for ensuing development or other activities

Speciation   The process whereby species multiply; the acquisition of reproductive isolation between populations

Speciational evolution   See Punctuationism

Species (biological)   A reproductively isolated aggregate of populations which can interbreed with one another because they share the same isolating mechanisms. See Evolutionary species concept; Nominalist species concept; Typological species concept

Species replacement   The turnover of species in time, owing to the extinction of some species and the origin of new species that replace them. Also referred to as species selection

Sport   In animal breeding, an aberrant individual produced by a major mutation

Stochastic processes   Processes consisting of a series of steps, each of which is random in direction. See Determinism

Sympatric speciation   The splitting of a species into two reproductively isolated species within the same area

Sympatric   Coexisting at the same locality; said of a population in breeding condition that exists within the cruising range of individuals that belong to a different species

Systematics   The science which studies the diversity of organisms

Taxon   A monophyletic group of organisms that share a definite set of characters and are considered sufficiently distinct to be worthy of a formal name

Taxonomy   The theory and practice of classifying organisms

Teleology   The actual or only seeming existence of end-directed processes in nature, and their study. See Finalism

Teleomatic process A seemingly end-directed process that is strictly controlled by natural laws such as the law of gravity or the first law of thermodynamics

Teleonomic process A process or behavior that owes its goal-directedness to the operation of a program

Theism The belief in a personal god who is forever present and able to affect any natural process at any time

Transformational evolution Gradual change of an object in the course of time from one condition of existence into another; usually combined with a belief in change from "lower" to "higher," or from less perfect to more perfect. See Lamarckism; Saltational evolution; Variational evolution

Typological species concept Defines species on the basis of degree of difference. See Biological species concept; Evolutionary species concept; Nominalist species concept

Ultimate or evolutionary causation In evolutionary biology, the historical factors responsible for the properties of species and individuals, and more specifically for the composition of the genotype

Uniformitarianism A theory, particularly promoted by Charles Lyell, that all geological changes, no matter at what rate they occur, are gradual. See Catastrophism

Variant A member of a variable population

Variational evolution Darwin's concept of evolution, which states that change occurs in every generation through the production of a large amount of new genetic variation and through the survival ("selection") of a small percentage of the varied individuals to serve as the progenitors of a next generation. See Saltational evolution; Selection; Transformational evolution

Yarrell's Law A generalization proposed by William Yarrell that "the longer a character was in the blood of a race, the more fixed it would be"

Zygote A fertilized egg; the individual that results from the union of two gametes and their nuclei

ACKNOWLEDGMENTS

This analysis and evaluation of Darwin's thought is based not only on Darwin's own writings but also on the critical studies of a dedicated group of Darwin scholars. To all of them I am greatly indebted. Their number is too large for me to mention them here individually, but their names can be found in the References. At the risk of being unfair to the others, I want to thank especially Frederick Burkhardt and David Kohn for their never-failing willingness to give me suggestions and information.

Walter Borawski again indefatigably typed countless drafts of the manuscript and assisted me in the preparation of the References, Glossary, and Index. Howard Boyer, Senior Science Editor at Harvard University Press, encouraged and supported this project from its inception. Susan Wallace, also at the Press, improved the raw manuscript immeasurably by eliminating redundancy, arranging a more logical sequence of subject matter, pointing out gaps in my original manuscript, suggesting stylistic improvements, and in various other ways. I cannot express in words my deep gratitude to her.

Ernst Mayr

Index

Adams, M., 139
adaptation, 56, 115
adaptationist program, 115
adaptive radiations, 157
affinity, 22
Agassiz, L., 8, 16, 38, 100
Alexander, R., 155
altruism, 155
amphimixis, 124
animal breeders, 71, 81
anti-creationism, 141
Archaeopteryx, 159
archetypes, 55
"argument, one long," 94
Aristotle, 50
asexuality, advantage of, 123
autobiography (Darwin's), 70

Baconian method, 9
Baer, K. E. v., 53
balance of nature, 75
balanced selection, 151
barnacles, 6, 45
Barzun, J., 108
Bates, H. W., 8
Bateson, W., 46
Bauplan, 160
Baur, E., 133
bdelloid rotifers, 123
Beagle, 4, 69
Beckner, M., 66
behavior, social, 154
Bergson, H., 66
Bible, 13
Blacher, L. I., 119
Blumenbach, J. F., 16
Blyth, E., 86
Bonnet, C., 78

Bowler, P. J., 55, 60
Boyle, R., 52
branching, 21, 22
"bridgeless gaps," 41
Buch, L. v., 19, 32
Buffon, G. L., 12, 51
Burian, R., 91

Candolle, A. P. de, 78
causations: proximate, 53; ultimate, 53;
    simultaneous, 147
causes, final, 50
central dogma, 120
chance, 49, 66, 140; or necessity, 59
change, 153
character divergence, 33, 36
Chardin, T. de, 66
Chetverikov, S., 134
chorda, 24
Christian faith, 75
Churchill, F., 119, 121
classification, theory of, 2
cohesion, of genes, 160
common descent, 21, 22, 95, 100, 163
competition, 58, 78
consciousness, 25
constraints, 129; ideological, 38; devel-
    opmental, 116, 164
continental drift, 105
creationism, 15, 16, 17, 99
Creator, 49
culture, 63
Cuvier, G., 16, 22, 77

Darwin, Annie, 15
Darwin, Charles: as agnostic, 13–15;
    five theories of, 35; as philosopher,
    50, 102; papers of, 71

Darwin, Emma, 15
Darwin, Erasmus, 3
Darwinian revolution, 89, 107, 132; first, 12, 25; second, 89
Darwinism, 90; as anticreationism, 93; as evolutionism, 93; as anti-ideology, 96; as selectionism, 97; as variational evolution, 97; as creed of the Darwinians, 98; as a new world-view, 101; as a new methodology, 104; current meaning of, 107, 162, 164; stages of, 143
de Beer, G., 71
de Vries, H., 46
deism, 93
Derham, W., 52
Descartes, R., 51
Descent of Man, 7
design, 38, 52, 54
determinants, genetic, 125
determinism, 48
development, 145, 158, 162
diagram, branching, 22
dinosaurs, 65
diversification, 62
diversity, 163
Dobzhansky, Th., 134
domestication, 44
dominance, 63

earth, age of, 16
East, E., 133
Ehrenberg, C. G., 76
Eimer, T., 61
Eldredge, N., 137
electrophoresis, 152
environment, emancipation from, 63
equilibria, punctuated, 153
essentialism, 40
eukaryotes, 62
evolution: gradual, 18, 43; saltational, 42; transformational, 43; variational, 43, 163; unpredictable, 44; mechanism of, 68; as fact, 112, 162; geographical, 137; vertical, 137
evolutionary synthesis, 106, 132–140, 141, 146
evolutionism: vertical, 17, 20; horizontal, 20

exemplar, 106
experiment, Darwin's use of, 10
explanatory model, Darwin's, 73
externalism, 39
extinction, 16, 55
eye, evolution of, 58

female choice, 89, 117
finalism, 50–67
Fisher, R. A., 143, 151
FitzRoy, R., 4
Ford, E. B., 134
fossil record, 16, 55, 138
founder populations, 148

Galapagos, 5
Galton, F., 45, 119
game theory, 156
gene flow, 148
genetic code, 23, 163
genetic drift, 119
genetic program, 122
genetic variation, source of, 124
geneticists, 139
genetics, 132, 139
genotype: domains of, 158; cohesion of, 130, 160; congealing of, 160
Geoffroyism, 126
geographical speciation, 20
geology, 16
germ plasm, 122
germinal selection, 125, 126
Ghiselin, M. T., 6, 42, 79, 105
gill arches, 24
Gillespie, N. C., 15
Gillispie, C., 94
God, 51, 52, 56, 96
Goldschmidt, R., 46
Goodman, M., 159
Gordon, S., 85, 102
Gould, J., 5, 18, 32
Gould, S. J., 112, 137
gradualism, 17, 18, 44, 114, 163
Grant, V., 145
Gray, A., 59, 100
Greene, J. C., 96, 101
group, cultural, 156
group selection, 156, 157
Gruber, H. E., 43, 71, 80

Haeckel, E., 24, 60, 157
Haldare, J. B. S., 155
Hamilton, W. D., 154
Harris, H., 152
Hartmann, E. v., 53
Henslow, J. S., 55
Herbert, S., 79
Herder, J. G., 52
Herschel, J., 9, 48, 56
Hessen, B., 39
higgledy-piggledy, theory of the, 49
Himmelfarb, G., 108
Hodge, J., 69, 95
Holbach, Baron, 51
holism, 138
homeoboxes, 158
homoplasy, 151
Hooker, J., 7
horizontal evolution, 20, 137
Hubby, J. L., 152
Hull, D. L., 14, 91, 106
Hume, D., 54
Huxley, A., 47
Huxley, J., 134
Huxley, T. H., 8, 24, 46, 63

ideologies, 38; opposing Darwinism, 40
inclusive fitness, 156
individual, 80; uniqueness of, 42
inductionism, 127
inductivism, 104
inference, 48, 108
inheritance: soft, 74, 108; of acquired characters, 74, 110, 118; blending, 109, 123; particulate, 129, 135
internal causation, 69

Jacob, F., 149
Jenkin, F., 46
Journal of Researches, 4
Jukes, T., 152

Kant, I., 50, 54
Kellogg, V. L., 60
Kimura, M., 152
kin selection, 156
King, J., 152
Kitcher, P., 91, 98, 105

Kleinenberg, N., 161
Kölliker, A., 46, 59
Kohn, D., 14, 15, 44, 69, 70, 73, 108
Kuhn, T., 132

Lamarck, J. B., 12, 16, 17, 19, 21, 43, 54, 62
Lamarckism, 120
laws, 49, 56, 57, 93
Leibniz, G. W., 45, 52
Lennox, J. G., 53
Lenoir, T., 50
Lewontin, R. C., 43, 152
Limoges, C., 71
Linnaean hierarchy, 23
Linnaeus, C., 20, 78
Löw, R., 53
Lovejoy, A., 44
lungfishes, 161
Lwoff, A., 139
Lyell, C., 4, 12, 16, 17, 19, 38, 41, 43, 55, 64, 101
Lysenko, 139

macroevolution, 138, 161
Malthus, R., 57, 69, 81, 84–85, 102
man: position of, 24, 38–39; origin of, 25
matter in motion, 96
Maynard Smith, J., 156
mechanists, 50
Mendel, G., 109
Mendelians, 46, 136
Mendel's laws, 149
method, scientific, 9
methodology, 104–105
Mill, J. S., 9, 41, 56
mind, 25
mockingbirds, 5, 18, 20, 22, 32
molecular biology, 149
molecular clock, 159
Monod, J., 66
Moore, J. R., 15, 59
Morgan, T. H., 60, 133
mosaic evolution, 130, 155, 158
Müller, F., 8
Muller, H. J., 151
mutation, spontaneous, 124
mutations, small, 133

Nagel, E., 66
natural kinds, 27, 41
natural selection, 57, 59, 65, 70, 86,
    113, 139, 164; path to, 68; cause of,
    86; validity of, 121; triumph of, 139.
    See also Selection
natural theology, 38, 52, 55, 68
naturalist, Darwin as, 3, 10, 139
naturalists, contributions by, 137–139
neo-Darwinism, 108
neo-Lamarckism, 60
nervous system, central, 63
neutral evolution, 151
notebooks (Darwin), 83–84
novelties, evolutionary, 138

organs, vestigial, 57
origin of life, 24
orthogenesis, 53, 60, 113
Osborn, H. F., 43
Ospovat, D., 15, 56
Owen, R., 22
Oxford, 139

pangenesis, 36
Patagonia, 4
pathogens, 87
perfecting principle, 61
perfection, 52, 55, 58; progression to-
    ward, 56
phenotype, 137
philosophers, 48, 50
philosophy: of science, 2, 9; hypothe-
    tico-deductive, 128
phylogeny, 112
physicalism, 48
physicists, 48
Plato, 40
pluralism, 105, 106, 147
population stability, 76
population thinking, 40, 42, 46, 74, 80,
    138
populations: local, 27; exponential in-
    crease of, 75
prediction, 48
principle of exclusion, 116
probabilism, 49
processes: stochastic, 49; teleomatic, 67
program: genetic, 158; somatic, 161

progress, 1, 53, 57; evolutionary, 62, 64
progressionists, 43
progressiveness, 63
punctuationism, 153

quality, 125
quinarians, 22

rate of evolution, 47, 159
Ray, J., 20, 52
recapitulation, 112, 145, 157
Recker, D., 95, 99, 101, 104
recombination, genetic, 122, 124
rectilinearity, 61, 65
reduction division, 129
religion, Darwin's, 13
Rensch, B., 134
reproduction, sexual, 82
resources, limitation'of, 76
revolution: first Darwinian, 12, 25; sec-
    ond Darwinian, 89
revolutions: intellectual, 1; scientific, 39
Rhea, 18
Richards, R. J., 15
Romanes, G. J., 110
Ruse, M., 13, 56, 82
Russia, 139

Sachs, J., 61
saltation, 18, 46
saltationism, 19, 42, 113
Sapp, J., 136
scala naturae, 21, 61
Schindewolf, O., 46
Schleiden, M., 55
Sedgwick, A., 55
selection: artificial, 83; of 87; for, 88;
    sexual, 88, 117, 161; evidence for,
    114; normalizing, 116; target of,
    117; germinal, 125. *See also* Natural
    selection
sex, significance of, 122, 124
sexual selection, 88, 117, 156, 164
sibling species, 27, 148, 161
Sigwart, C., 66
Simpson, G. G., 28, 65, 135, 159
Smith, A., 102
Smith, S., 71
sociobiology, 154–157

socioeconomic factors, in scientific revolutions, 39
somatic programs, 158
soul, in man, 39
special creation, 102
speciation, 5, 18, 19; allopatric, 20; defined, 31–34; gradual, 32; on islands, 32; sympatric, 32
species concepts, 26; nominalist, 27; typological, 27; evolutionary, 28; biological, 29; Darwin's, 30
species replacement, 88
"species book," 6
species: constancy of, 12; defined, 27–30; in plants, 29; multiplication of, 31; incipient, 33
species selection, 64, 88
species succession, 64
Spencer, H., 102, 125
Stanley, S. M., 43
stasis, 64, 65
Stebbins, G. L., 145
stochastic processes, 140
struggle for existence, 77
Sulloway, F. J., 5, 18, 34
synthesis, evolutionary, 106, 132–140, 141, 146
systematics, 146
systems, 97; adapted, 67

target of selection, 68
teleology, 50, 66; cosmic, 53, 67
teleomatic processes, 67
teleonomic processes, 67
theism, 12
Toulmin, S., 53
transmutation, 43
Trivers, R., 155

twin studies, 154
typological thinking, 42

uniformitarianism, 16, 43
uniqueness, 74
use and disuse, 36, 109, 120

van Beneden, E., 123
variation, 42, 109; accidental, 49; source of, 118; qantitative, 125; gradual, 136
variety, 29
vertical evolution, 17, 20, 137
Voltaire, F. M. Arouet, 54

Waddington, C. H., 157
Wagner, M., 19, 32
Wallace, A. R., 6, 19, 73, 90, 92, 97
Wedgwood, E., 6
Wedgwood, J., 3
Wegener, A., 12
Weismann, A., 60, 110, 108–131; as panselectionist, 116; criticism of, 127; contributions of, 128
Westoll, S., 161
Whewell, W., 9, 41, 56
White, M. J. D., 145
Whitehead, A. N., 66
Whitman, C. O., 61
"why" questions, 4
Wiley, E. O., 28
Willis, J. C., 46
Willmann, R., 28
Wilson, A. C., 159
Wilson, E. O., 154

Yarrell's Law, 44

Zeitgeist, 1, 40

Library of Congress Cataloging-in-Publication Data

Mayr. Ernst. 1904–
One long argument : Charles Darwin and the genesis of modern
evolutionary thought / Ernst Mayr.
p. cm.
Includes bibliographical references and index.
ISBN 0-674-63905-7
1. Evolution. 2. Evolution—Philosophy.
3. Darwin, Charles, 1809–1882. I. Title.
QH371.M336   1991
575—dc20

91-11051
CIP